ALSO EDITED BY BORA ZIVKOVIC

The Open Laboratory: The Best Writing on Science Blogs 2006

The Open Laboratory: The Best Science Writing on Blogs 2007
(Reed A. Cartwright, guest editor)

The Open Laboratory: The Best in Science Writing on Blogs: The 2008 Edition
(Jennifer Rohn, guest editor)

The Open Laboratory: The Best in Science Writing on Blogs: The 2009 Edition
(Scicurious, guest editor)

The Open Laboratory: The Best of Science Writing on the Web 2010
(Jason G. Goldman, guest editor)

THE BEST SCIENCE WRITING ONLINE 2012

BORA ZIVKOVIC, SERIES EDITOR

JENNIFER OUELLETTE, GUEST EDITOR

SCIENTIFIC AMERICAN / FARRAR, STRAUS AND GIROUX

NEW YORK

THE BEST SCIENCE WRITING ONLINE 2012

Scientific American / Farrar, Straus and Giroux
18 West 18th Street, New York 10011

Distributed in Canada by D&M Publishers, Inc.
Printed in the United States of America
First edition, 2012

Owing to limitations of space, all acknowledgments for permission to reprint these essays appear on pages 325–28.

ISSN 2168-0299
ISBN 978-0-374-53334-2

Designed by Jonathan D. Lippincott

www.fsgbooks.com
books.scientificamerican.com

10 9 8 7 6 5 4 3 2 1

CONTENTS

PREFACE

It was a gorgeous December day in 2006 when I met with my friend and colleague Anton Zuiker at the 3 Cups coffee shop in Chapel Hill to discuss plans for the inaugural Triangle Science Blogging Conference, a gathering we hoped would introduce local scientists, bloggers, teachers, and programmers to one another and start a dialogue and potentially some projects for promoting science on the Web. One of the ideas was to put together an anthology of the best science blog posts of the year and have Lulu.com, as a conference sponsor, publish it and give a copy to every attendee at the conference. Seeing as there were only four weeks left before the meeting was to start, Anton thought there was not enough time to put together such an ambitious project. I begged to differ.

I posted a call for submissions on my blog that afternoon, and over the next week or so, bloggers and their readers nominated several hundred blog posts. I asked a dozen or so friends in the science blogosphere to help me read all the entries and select the fifty pieces we thought were the best representation of the quality of writing we envisioned for the collection. They also helped with the technical aspects and design decisions related to the book.

The Open Laboratory 2006—the title of which was also chosen in an informal online discussion—was published just four weeks after our first conversation about it, and the science blogging community instantly fell in love with the project. It soon became apparent that this collection couldn't be a one-off—the collaborative effort showed us how excitement about the project unified and energized the community. I was also reminded of the power of the World Wide Web to draw people together and organize them for communal action, not to mention that we could see, highly concentrated in one place, incredible examples of the amazing writing happening online. The book would have to be done on an annual basis.

Knowing how much work goes into putting an anthology together, I decided to ask for help. Each year I selected a prominent science blogger to serve as the guest editor. Reed Cartwright (of the blog *Panda's Thumb*), Jennifer Rohn (the editor of LabLit.com and author of two books, *Experimental Heart* and *The Honest Look*), Scicurious (popular for her work on multiple blogs: *The Scicurious Brain*, *Neurotic Physiology*, and *Neurotopia v 2.0*), and Jason Goldman (of the blog *The Thoughtful Animal*) helped bring the individual collections published since 2006 into the world, with each of them adding a different flavor and personal touch that made each year delicious in its own way.

In addition, over the last couple of years Blake Stacey had handled all the technical aspects of book formatting, and David Ng and Glendon Mellow had collaborated on the cover design. Everyone involved in the collections added something new to the project and improved the process. This year, I was looking for a guest editor with feet in both camps: a prominent science blogger who was also an experienced author. And there was really only one person in my mind who fit my criteria perfectly: Jennifer Ouellette. I had known Jennifer for a long time and knew that she would bring in an impeccable taste for writing, long editorial experience in the book world, and a sense of humor that would make the hectic job of corralling fifty wild and woolly bloggers a pleasure instead of a pain.

The main appeal of the *Open Laboratory* anthology to the bloggers is that it is a community-based project, and entirely transparent in its execution. The entries come in throughout the course of the year, and the complete list is made public every week. Smaller or newer bloggers thus regularly get new incoming links and new readers. Bloggers with entries included in the collections proudly sport the buttons on the sidebars of their blogs. Those science bloggers who are also professional writers list their inclusion in the anthology as a publication on their résumés. Bloggers who are active researchers list it in the "outreach" sections of their CVs.

But there is something more to it than just how much bloggers love this book. It is seen as a bridge between the online and offline worlds. Everyone involved buys extra copies to give to friends and relatives who are not as Web-savvy and may not realize what amazing writing transpires on science blogs. The hope is that the book will open the wonderful world of science blogs to a broader audience that may not have been aware of it before—which is why we felt that, in order to reach as broad an audience as possible, we needed to move the project to a new level of professionalism. So this year, the collection finds a home at the Scientific American imprint at Farrar, Straus and Giroux. I spoke with Scientific American / Farrar, Straus and Giroux senior editor Amanda Moon, and she shared a similar vision for this collection—to find ways to bring the online world and the print world together, exploring the cutting edge of book publishing and supporting the online science writing community at the same time. It was important to us that the submission process, the weekly publicizing of all the submitted links, the reviewing process, and the announcement of the final list of pieces that made it into the book be communicated as openly as before, and that the community spirit continue to flourish.

I hope you find the articles in this collection, now entitled *The Best Science Writing Online 2012*, as riveting as I do. Both in terms of production and in terms of the quality of writing produced by the rapidly expanding and maturing science blogosphere, this is the best collection yet. I hope you enjoy reading it as much as we enjoyed putting it together.

Bora Zivkovic, Series Editor
Chapel Hill, North Carolina, January 2012

INTRODUCTION
Blogs Are Dead! Long Live the Blogs!

In February 2011, a headline in *The New York Times*[1] mourned the anticipated demise of blogs, because young people—a Pew study had shown—were flocking to social networking platforms instead. It was not the first time such a pronouncement had been made, nor is it likely to be the last. Just a few weeks earlier, *The New York Observer* had run the headline "The End of Blogging,"[2] and by December of that same year, columnists were wringing their hands over the end of tech blogging's "Golden Age."[3]

Blogging is dead? Really? Again? Someone should inform the science bloggers, because you'd never know this was the case from the cornucopia of riches featured in this year's *Best Science Writing Online* anthology. Rather, I am struck anew by the sheer diversity in voice, style, subject matter, and creativity that one finds across the blogosphere, and the science blogosphere in particular.

There is poetry. Savvy reportage and critical analysis of new scientific papers. In-depth profiles. Personal reflections. Humor. Thoughtful commentary on science and social issues. Careful explication of complex scientific concepts written in accessible language. And yes, there are long-form features and investigative journalism. Above all, there are stories—drawn from history, popular culture, the laboratory, and personal experiences.

In these pages, you'll find essays on fluid flow, airplane turbulence, poisons, the last space shuttle launch, and why hot water freezes before cold. You'll learn about botulism in fermented seal flipper, the microbiology of beards, what it's like to be a twin, and how geology can help solve murders. Indulge in cheeky ruminations on sustainable seafood, menstruation, why women cry, and how to take the real measure of a man. And did I mention the pirates? Oh yes, we have pirates.

Needless to say, winnowing 720 entries down to 50 (plus one poem) was no easy feat, given the quality of so many of the submissions. I absolutely could not have done it without the help of all the volunteer reviewers. Even so, series editor Bora Zivkovic and I engaged in much mutual hand-wringing, particularly over the last dozen or so cuts we were forced to make.

Our criteria were simple: Was the post substantive? Was the post factually accurate? Did the post have an interesting and unique perspective? Was the writing of high quality? Was the "voice" unique and compelling?

As this year's anthology editor, I was also looking for other qualities. Did the post surprise me, delight me, or just plain move me? Did the post take risks or move out of the writer's comfort zone? Ultimately, I looked for good storytelling and for posts that showed the human element shining through the science—because that personal touch is what makes blogging such a powerful communication medium.

I think we ended up with a good mix, showcasing scientists and science writers (both established ones and fresher faces) across a broad range of fields, not to mention an almost perfect fifty-fifty split between male and female bloggers, although that wasn't by design. Each puts his or her unique stamp on this dynamic format.

These selections offer just a sampling of the rich variety in a thriving science blogosphere that continues to evolve and reinvent itself to adapt to the changing media landscape. It's time to toss out the antiquated notion that "science blogging" is somehow distinct—with a faint whiff of illegitimacy, even—from "real" science writing, at a time when "legitimate" science writing is moving increasingly online and every major science magazine, it seems, is building up its own stable of bloggers.

Far from becoming irrelevant, science blogging has emerged as an essential activity for science writers as we find ourselves with a professional presence on Twitter, Facebook, and Google+, not to mention "microblogging" platforms like Tumblr. And it's become an equally essential tool for scientists themselves to connect and communicate directly with the general public.

As Bora has said (repeatedly): "Blog is software." Nothing more. What you do with that software is entirely up to you. And based on this year's round of essays, I, for one, see great things ahead in the future of the science blogosphere.

Jennifer Ouellette, Guest Editor
Los Angeles, California, 2012

NOTES

1. Verne G. Kopytoff, "Blogs Wane as the Young Drift to Sites Like Twitter," *New York Times*, February 20, 2011. www.nytimes.com/2011/02/21/technology/internet/21blog.html.
2. Dan Duray, "The End of Blogging," *New York Observer*, February 1, 2011. www.observer.com/2011/tech/end-blogging.
3. Jeremiah Owyang, "End of an Era: The Golden Age of Tech Blogging Is Over," *Web Strategy* (blog), December 27, 2011. www.web-strategist.com/blog/2011/12/27/end-of-an-era-the-golden-age-of-tech-blogging-is-over/.

THE BEST SCIENCE WRITING ONLINE 2012

I

SOMATA

AUBREY J. SANDERS

I was born a body of worlds
a carnal web of cosmic pearl
billions of stars that hold me to my bones,
and when one day their cores collapse
I will shed my skin in ash
and sleep among the mosses and the stone.

I'll grow into the vine that licks the ruin
writhe beneath the savage moon
my scattered cinders eaten at the roots,
and when the ravaged willow moans again
she will take me in her veins
and shake me from her hair an astral fruit.

For we forgot a fact that we once knew,
the only ancient truth,
the knowledge of our primal origin:
That from the feral night we came as dust
born from stellar wanderlust
and unto the stars we will return again.

AUBREY SANDERS will graduate from the George Washington University in May 2012 with a B.A. in English and creative writing. She currently interns at the Grosvenor Literary Agency in Bethesda, Maryland, and is working on her first novel. She believes in the poetry of the cosmos and will be an astronomer in her next life.

2

MAKE HISTORY, NOT VITAMIN C

EVA AMSEN

"Does the Flap of a Butterfly's Wings in Brazil Set Off a Tornado in Texas?"
—Edward Lorenz

This is a story about a tiny molecular shift affecting war, politics, disease, agriculture, and international corporations. Like all good stories, it also contains a healthy dose of biochemistry and genetics, some pirates, and a few rodents of unusual size. The very start—the event that set everything in motion—is a genetic mutation that happened millions of years ago, but we'll get to that. First, let's meet the pirates.

The pirates in this story are Dutch, and they were active near the end of the sixteenth century. During this time, the Netherlands were occupied by Spain, and after a period of repression, the northern (Protestant) provinces started to fight off the Spanish. They were most successful on water. From 1568 onward, several ships received government permission to attack and plunder Spanish ships. These *Watergeuzen* dominated at sea, but in 1572 they captured the city of Brielle, marking a turning point in the Eighty Years' War.

Meanwhile, a large part of the income for the Spanish side of the war came from trade with the East Indies. The European supply of pepper was provided solely by Portuguese fleets, and the trading post in Lisbon was no longer easily accessible to the Dutch while they were at war with Spain. Pepper was extremely valuable in those days, and the Portuguese kept their routes secret to make sure nobody else would cash in on the spice. But eventually, Dutch ships found a route to the East

Indies. They sailed south, all the way around Africa, and returned with enough spices to finally make some money.

Finding a trade route to the East Indies led to the formation of the Dutch East India Company (VOC) in 1602—the first multinational corporation, and the first company to sell stock. The company did more than buy and sell spices. For several years, it had a monopoly on colonial activities in Asia, and it had the power to take prisoners and establish colonies. During its existence, the VOC boosted the economy of the Netherlands to the top of the world. This period of economic growth is referred to as the "Golden Age" in Dutch history.

Money may not buy happiness, but the sudden wealth of the country certainly formed the perfect environment to nurture artistic endeavors and encourage major scientific progress. These were the century and country in which Rembrandt painted *The Night Watch* and Antonie van Leeuwenhoek developed the microscopes with which he first observed single-celled organisms. The effects of the VOC trade have shaped entire fields of art and science, all because a few ships found a route to the East Indies in a time of economic need.

There was just one problem with the VOC trade route to the East Indies: It was quite long.

Scientific progress notwithstanding, there was no suitable way to keep the crew's food, especially fresh fruit and vegetables, from going bad before they were even halfway there. This was a problem, because without fresh fruit, the crew was prone to scurvy. Scurvy had been the scourge of sea travelers since the fifteenth century, when ships started to sail across oceans and stayed away from home—and fruit—for too long. Starting with some spots on the skin, scurvy progressed to bleeding from mucous membranes, ulcers, seeping wounds, loss of teeth, and eventually death. Fifteenth-century explorers could lose up to 80 percent of their crew to scurvy. The solution was known and simple: eat lots of fresh fruit.

Scurvy is caused by a lack of ascorbic acid, better known as vitamin C. Our bodies use this vitamin for many metabolic processes, such as producing collagen and repairing tissue damage. Without vitamin C, we essentially slowly start to fall apart: skin breaks open, wounds won't heal, teeth fall out.

But we humans are one of the few animals that need to eat fruit and vegetables to keep our vitamin C levels up. Most animals are quite capable of synthesizing their own vitamin C. Most, but not all. We share our need for fruit and veggies with other primates, including closely related apes as well as monkeys and tarsiers.

Our inability to synthesize vitamin C is the result of a mutation that occurred more than 40 million years ago in our shared primate ancestor, affecting the gene that encodes the L-gulonolactone oxidase (GULO) enzyme. Normally, this enzyme catalyzes a crucial step in the formation of vitamin C. But in humans and related primates the genetic mutation produces a broken enzyme. It doesn't work, and we can't make our own vitamin C anymore. Luckily, it's quite easy to compensate for the lack of GULO by simply taking in vitamin C via our diets, but this also means that there was no selective pressure for a functional GULO, and we primates have been living with a broken version ever since.

The relative ease with which animals can compensate for no longer producing their own vitamin C is illustrated by the fact that the mutation that disabled our GULO enzyme millions of years ago was not the only mutation in the animal kingdom to shut down vitamin C biosynthesis. It happened at least three other times: bats, guinea pigs, and sparrows also have defective GULO enzymes and get vitamin C via their diets.

The mutation in the guinea pig's ancestor happened more recently than ours—possibly "only" about 20 million years ago—but that is still far enough back to also have affected another member of the Caviidae family. The capybara also needs a steady diet of vitamin C to keep a hold on its title of largest living rodent on Earth. Especially in captivity these R.O.U.S. (rodents of unusual size) are, like the sailors and pirates of yore, at risk of scurvy unless they eat enough fresh vegetables.

Speaking of fresh vegetables, how were the VOC crew going to manage the journey to the East Indies, which took longer than the expiration date on their perishables? The ideal solution was to restock along the way, but the continent of Africa was not exactly a farmers' market where you could just get some more fruit and veg when you needed it. Well then, they would just have to make a farmers' market. The VOC took several Dutch farmers and settled them in South Africa to grow more food for the ships passing by along their trade route. The restocked ships could then sail on with a scurvy-free crew.

If the VOC crew had been able to make their own vitamin C, like most animals do, they wouldn't have had to bring farmers to South

Africa. That move, guided by a mutation that happened millions of years ago, entirely shaped the more recent history of South Africa. How? Here's a hint: the Dutch word for farmer is "boer."

The Boer population of South Africa were the direct descendants of the farmers relocated there to supply the VOC ships with the fruit and vegetables for their voyage to and from the East Indies. After the VOC was disbanded and British colonials settled in South Africa, the Boer population moved away from the Cape. Conflicts between the Boers and the British Empire, most notably the Second Anglo-Boer War at the end of the nineteenth century, directly led to the formation of the Union of South Africa in 1910, which was the predecessor of the current-day Republic of South Africa.

So there you have it. In a scene set by pirates, and with R.O.U.S. lurking in the background, an entire country, with all its political and cultural complications, was formed as a result of a method to distribute fruit and vegetables to the crew of seventeenth- and eighteenth-century trade ships to compensate for a genetic mutation that makes humans incapable of synthesizing their own vitamin C.

Our broken GULO enzyme may not have been able to make vitamin C for millions of years, but it's made history just fine.

EVA AMSEN is a former biochemist who left the lab for the laptop and now spends most of her time communicating with researchers, reading about science and publishing, interviewing scientists about their hobbies, and maintaining far too many blogs. Most of Eva's online presences can be found at http://easternblot.net.

3

SAVING ETHIOPIA'S "CHURCH FORESTS"

T. DELENE BEELAND

Historically, fundamentalist Christians believed mankind had a God-given right to use Earth and all its resources to meet humanity's needs. After all, Genesis gave man dominion over Earth. In modern times, green messages of sustainability are permeating some Christian groups, yet on the whole sustainability remains more of a commitment on paper than in practice. But across the Atlantic, a much different attitude prevails. Ethiopia has the longest continuous tradition of Christianity of any African country, and followers of the Ethiopian Orthodox Tewahedo Church believe they should maintain a home for all of God's creatures around their places of worship. The result? Forests ringing churches.

There are some 35,000 church forests in Ethiopia, ranging in size from a few acres to more than 700 acres. Some churches and their forests may date back to the fourth century, and all are remnants of Ethiopia's historic Afromontane forests. To their followers, they are a sacred symbol of the garden of Eden—to be loved and cared for, but not worshipped.

Most church forests are concentrated in the northern reaches of the country, especially in the Lake Tana area. Here, most of the Afromontane forests have been cut down to make clearings for agriculture, pastures for livestock, and settlements. It is said that if a traveler to the area spies a forest, it surely has a church in the middle. Many also have freshwater springs.

These spiritually protected woods, also known as Coptic forests, comprise a decent chunk of the 5 percent of Ethiopia's historical forests that are still standing. Massive deforestation has rendered these church forests true islands—green oases peppering a land laid bare.

(Courtesy of D. M. Jarzen, Ph.D.)

"Connecting them to other forests is a luxury we can't even consider," says international tropical ecologist and researcher Margaret Lowman, because the lands between them are predominantly crop fields today. Fondly known to her colleagues as Canopy Meg, Lowman is a canopy researcher who has studied forests on five continents.

What remains is fragile and isolated by habitat fragmentation. Conservationists have made these church forests a centerpiece of the country's fight to retain its biodiversity. These beacon-like green swaths have become refuges for all kinds of species—but no one really knows what is at stake because they are extremely poorly studied.

Awakening Awareness

Alemayehu Wassie Eshete is an Ethiopian forest researcher who did his Ph.D. work on his country's Coptic forests. His dissertation and a few related papers may form the entirety of the published scientific literature on Ethiopia's church forests, according to Lowman.

She met Wassie at a scientific meeting a few years ago. When she asked him what he was going to do next on Coptic forests, he wept in frustration. With very limited economic means and few international connections, Wassie felt he needed help with visibility and expertise to study—and conserve—his country's forests. Lowman began helping him, and in 2009 he invited her to visit the forests in person. She went.

With Wassie's help, Lowman gave a PowerPoint presentation to about 100 men, mostly priests. Some had traveled for days to attend the meeting. She showed them images of their church forests.

Though some had never seen a computer, they recognized their forests, and she says they gasped audibly at the pictures showing the woods shrinking inward over time. Lowman says they intuitively grasped the value of ecosystem services because, for them, it fell under the purview of their spiritual duties to protect the biodiversity around their places of worship.

Findings from Terhi Evinen's master's research in forest sciences at the University of Helsinki revealed in detail how the priests conceive of their spiritual forests. "What [matters to them] religiously is the number of trees, not the ecological health of the forests," Evinen wrote to me. "The trees are said to be the jewelry of the church and the more trees a church has the more appreciated it is since the tree canopy prevents the prayers from being lost to the sky."

Despite this, Evinen and Lowman agree the biggest threats to these forests are not external factors such as industrial loggers or agribusiness. Rather, the biggest threat lies inside: the church members and clergy who use the forests for firewood and rely on them for livelihoods.

The clergy and church members use the wood from trees to repair their church, to make charcoal for church activities, and to make sacred utensils. Plants in the forest are eaten or used to make dyes. Deadfall is sold to congregants for cash.

The trip ingrained in Lowman's mind that engaging the priests and church members was a vital part of studying and conserving the remaining Coptic forests.

Insightful Insects

In August 2010, Lowman returned to Ethiopia and led a team of thirteen scientists in surveying several church forests on the south shore of

Lake Tana and at a rural village about sixty miles to the northeast, Debre Tabor. The purpose of the team's trip, one hopes the first of many, was to assess the insect biodiversity and the economic importance of the tree species remaining in these sacred forests.

"We didn't even know if birds or mammals would be left in these highly threatened areas," Lowman says. "We chose insects to survey because we wanted an index of something we hoped would be healthy. It would be awful to give the local people bad news right at the start. Mammals were worrisome because it may be they've all been hunted or poached. Birds were also problematic because the fragmentation level is so high. But also, pollinators are such an important part of ecosystem services that we thought being able to educate the local people about them at the same time would be useful."

And this embodies the heart of how Lowman has conducted her scientific career: equal parts science and public outreach.

Lowman's team examined every level of the forests: they climbed into the canopies, shook bugs from tree limbs, and set insect traps.

They surveyed the lower reaches and scoured the forest floor where dung beetles carry out important waste-removal processes. They surveyed ants, beetles, flies, and wasps, and counted birds, mostly in the surrounding fields. The team collected in the day and during the night, rain or shine. By the end, they had gathered data on 5,500 insects, mostly from two church forests: Debresena, outside of Debre Tabor; and Zahara, close to Bahir Dar.

A month before the research trip, Lowman started a new job as the director of the Nature Research Center at the North Carolina Museum of Natural Sciences. Part of the center's mission is to communicate science to the public and actively engage children. So she live-streamed video of the fieldwork back to the museum as a pilot project. Soon the researchers garnered the attention of local children, who watched and even helped the researchers collect their specimens. Lowman described it like this in *North Carolina Naturalist*: "Armed with our nets and ropes and vials, we attracted a large swath of children who watched our every move and marveled at the six-legged creatures swept from the foliage. Despite the language barriers, we all laughed when ants fell on our heads, and shrieked with joy when a purple beetle appeared on the surface of our collecting tray."

When she returns next year, Lowman plans to further engage the children through educational activities to teach them about the importance of their remaining forests. Though 95 percent of their forests have been wiped off the map, the remaining church forests provide important ecosystem services. They harbor pollinator species that forests and crops require, and they sequester carbon and conserve water. Wassie estimates that 1.1 billion tons of soil is eroded in northern Ethiopia every year. The trees of the church forests help to hold a small portion of the landscape's soil steadily in place.

Fences, Feces, and Farming

When people worship at the churches, it may be a several-hour to an all-day-long affair. Which means that people often need to relieve themselves in the forests, where there are typically no toilets. Priests and disciples also live in huts in the forest, using it for nature's call because no other facilities exist.

"We found a preponderance of dung beetles specific to human feces," Lowman says. "So obviously the system is being degraded even without intent." In fact, more than three-quarters of the faunal collections from the researchers' ground samples were these dung beetles.

When the team asked a priest what ideas he had for how the researchers could help them, he asked for toilets. Lowman says the original research project has now blossomed into fund-raising to help the communities build fences (to keep livestock out) and to build pit toilets. Her team calculated they need to raise $42,000 for latrines and $271,871 for fences and walls. Lowman holds hope that improving the hygiene of the people who use the churches will not only build trust but also enhance the health of the forests themselves.

Another opportunity Lowman sees for improving Ethiopia's church forests reflects back upon the force that created them: agriculture. If the farmers could get more productivity out of their crops, then they could use less land for agriculture. Some of the land could be reforested.

"But they need more food as the population increases. They are starving," Lowman says. "You can't expect hungry people to conserve forests when they need to plant an extra row of corn or millet."

Lowman retains her strongest optimism for the youth of these areas.

This is not surprising, since she has devoted a large part of her career to outreach geared to schoolkids.

In the case of Ethiopia's precious green dots, Lowman's research team is slated to head back during the dry season of 2012, and educational outreach to children ranks as high on her priority list as surveying insects and trees.

David Jarzen, a palynologist with the Cleveland Museum of Natural History, was formerly at the Florida Museum of Natural History, where he was a member of the research team led by Lowman in 2011. He summarized the fate of Ethiopia's Coptic forests in an article he wrote with his wife, Susan, and Lowman for the Palynological Society's newsletter:

> The future of these church forests remains to be seen. Some of the team scientists are already planning a return visit to the region during the dry season in January or February. A closer, more detailed study of the fauna and flora, over several years, covering wet and dry seasonal changes of these forests is needed to fully understand the nature of the forests and strategies for their conservation. Whether these people allow their forests to remain intact, or cut into small plots as they are now will determine the fate of their culture.

Which may be why Canopy Meg is investing in Ethiopia's kids to help save their country's forests.

REFERENCES

Eshete, Alemayehu Wassie. "Ethiopian Church Forests: Their Contribution in Combating Ecological Degradation and Climate Change Effects in Northern Ethiopia." Society for Ecological Restoration, 2005. PowerPoint presentation. www.ser.org/files/ppt/Alemayehu%20Wassie%20Eshete%20PPT.pdf.

Jarzen, David M., Margaret D. Lowman, and Susan A. Jarzen. "Ethiopia: A Land in Need." T.R.E.E. Foundation, 2010. PowerPoint presentation. http://canopymeg.com/wp/2010/11/02/ethiopia-a-land-in-need/.

Jarzen, David M., Susan A. Jarzen, and Margaret D. Lowman. "In and Out of Africa." AASP—*The Palynological Society Newsletter* 43, no. 4 (2010): 11–14. www.palynology.org/pub.html#nl.

Lowman, Margaret D. "CSI in Ethiopia: Children Survey Insects." *North Carolina Naturalist* (Fall/Winter 2010–11). http://naturesearch.org/2011/01/csi-in-ethiopia-children-survey-insects/.

——. "Finding Sanctuary: Saving the Biodiversity of Ethiopia, One Church Forest at a Time." *Explorers Journal* 88, no. 4 (2010–11): 26–31.

VIDEOS

Canopy insect trapping: www.youtube.com/watch?feature=player_embedded&v=KbI9G9na4pY

Comparing insects in the fields and the church forests: www.youtube.com/watch?feature=player_embedded&v=oclr7gSd3qo

Debresena Church: www.youtube.com/watch?feature=player_embedded&v=sxke32uImgY

T. DeLENE BEELAND is an independent writer and the author of *The Secret World of Red Wolves: A True Story of North America's Other Wolf*, forthcoming in 2013 from the University of North Carolina Press. She enjoys writing about human interactions with nature and ecology. Beeland lives in western North Carolina, where she relishes riding her beloved road bike along the Blue Ridge Parkway.

4

WHAT IT FEELS LIKE FOR A SPERM

AATISH BHATIA

In a delightful 1977 paper entitled "Life at Low Reynolds Number,"[1] the physicist Edward Purcell calculated that if you a push a bacterium and then let go, it will coast for a distance equal to one-tenth the diameter of a hydrogen atom before coming to a stop. And it will do this in under a millionth of a second. Bacteria clearly inhabit a world where inertia is utterly irrelevant. And it's not just bacteria that are thwarted in this way. The same holds true for sperm, or any other microscopic creature in the business of getting around. It has to do with being really small. Shrink down to the size of a sperm, and you'll find yourself in an alien world where the very laws of physics seem to conspire against you.

This is strikingly different from the way a fish can slice through water. So why does size matter so much for a swimmer? And how do sperm ever manage to get anywhere? What makes the world of a sperm so fundamentally different from that of a sperm whale? To answer these burning questions, we need to dive into the physics of fluids.

We don't usually learn about the physics of squishy things. Physics textbooks deal with well-behaved objects like rigid blocks, billiard balls, and inclined planes. What these solid objects have in common is that they don't like to be deformed. If you bend a rubber eraser, it'll come back to its original shape. But the moment you snap an eraser or break a Slinky, you've stretched past the limits of this rigid world.

While the rigid universe is orderly and methodical, the squishy universe is an entirely different beast. Unlike solids, fluids such as liquids or gases almost seem to rejoice in their ability to deform. They take the shape of whatever you place them in, morphing forms with reckless abandon.

I was recently out paddling, and noticed that as you move the paddle through water, tiny whirlpools develop along its sides. These whirlpools grow in size, become self-sustaining, and break off and float away. Eventually they die out, as they lose their energy to the fluid around them.

You could also watch the spirals and vortices created by rising smoke. Or notice the strange shapes made by the wind as it sweeps through the clouds. It's as if fluids have a life of their own, often wondrous and beautiful, and other times surprising and counterintuitive.

But the motion of fluids is notoriously hard to predict. It's so difficult that if you can solve the equations of fluid flow, there are people willing to offer you a million dollars.

The difficulty comes from a mathematical property of the equations known as nonlinearity. Simply put, a nonlinear system is one where a small change can lead to a large effect. The same thing that makes these equations difficult to solve is also what makes fluids surprising and interesting. It's why the weather is so hard to predict—tiny changes in local temperatures and pressures can have a large effect.

At this point, most reasonable people would throw their arms up in despair. But physicists are an unreasonably persistent bunch, and when faced with an equation that they can't solve, they try to get some insight by looking at what happens at extremes. For example, thick and syrupy fluids like glycerine behave in a surprisingly orderly fashion.

Picture a flowing river. If there is an obstruction to the water's path, like a rock jutting out of the surface, the water will move around it and swirl back upstream. Behind the rock, the water remains relatively calm. What you get is a spot on a moving river where the water is remarkably still. These calm spots are called eddies, and kayakers treat them as parking spots on the river.

But fluids don't always behave like this. If you replace all the water in a river with a viscous fluid like glycerine, there won't be any eddies. The syrup will simply follow the contours of the rock and smoothly flow around it.

In one case we have orderly flow, and in the other, eddies and turbulence. So is there any way to know, in a given situation, what kind of

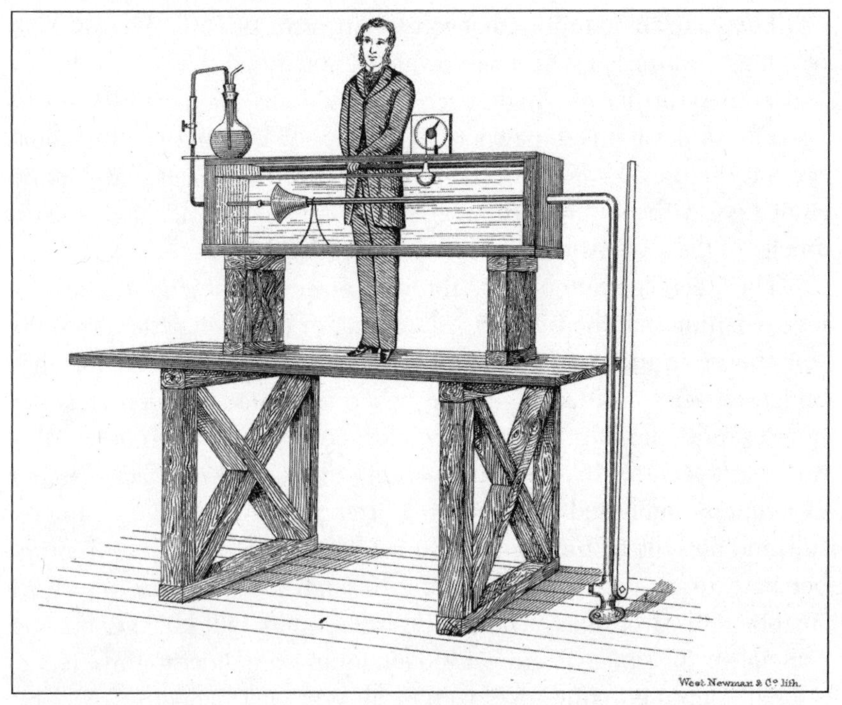

(Royal Society of London)

flow will result? This question was answered by the physicist Osborne Reynolds in 1883, and he answered it in style.[2]

Here is Reynolds's elegant experiment. He sent fluid flowing through a thin pipe (analogous to the river), and injected colored dye in a small section of the flow. He watched the dye flow down the tube, and could plainly see whether the flow was smooth or disorderly. By tweaking the parameters in this experiment, he was able to discover the conditions that ensure an orderly flow.

What he found is that there is one simple, magic number that can predict what is going to happen. It neatly ties together all the different physical quantities involved. It's been named the Reynolds number (Re for short), and is given by

$$Re = \frac{\text{density} \times \text{speed} \times \text{length}}{\text{viscosity}}$$

These are all quantities that you can measure directly. The viscosity of a fluid is a measure of how slowly it flows. Syrupy fluids like honey and corn syrup have a high viscosity, gases like air have a very low viscosity, and water is somewhere in between. The length in the equation on the previous page describes the size of the object that you are studying (say the width of the rock in our river). Reynolds used the diameter of the pipe. And the speed is that of the fluid.

The Reynolds number has the nice property of being dimensionless, meaning that the number is the same in whatever system of units you choose to measure the quantities (speed, on the other hand, has dimension—it's a different number when measured in kilometers per hour or miles per hour). What Reynolds found is that the moment this number exceeds 2,000, you suddenly get turbulent flow. In fact, a recent experiment[3] published in the journal *Science* verifies this surprising result, and puts the turning point at Re=2,040. The specifics of this number have to do with a fluid moving through a cylindrical tube with smooth walls. In different situations, the number will change, but the principle is the same. There is a sudden jump from order to turbulence.

The opposite figure gives you an idea of what happens as you increase the Reynolds number. Here's an analogy. The low Reynolds number world is like a collectivist ideal, where water moves along uniformly like soldiers marching in step. The high Reynolds number world is the individualist nightmare, where everyone looks out for themselves. Think of a march versus a mob.

We can arrive at this number from another route. There are two fundamentally different types of forces that act on an object immersed in a fluid. The first kind are inertial forces. This is like the push you give to the water when you take a stroke while swimming. Inertia is what allows water particles to keep moving undisturbed.

On the other hand, you have viscous forces, which measure the tendency for the fluid to smooth out any irregularities. To use the above analogy, inertial forces reflect the individuality of bits of fluid, and viscous forces are like a communist government enforcing conformity. When you take the ratio of these forces, you get back the Reynolds number.

This number is of immense importance to aeronautical engineers and to biologists interested in locomotion.

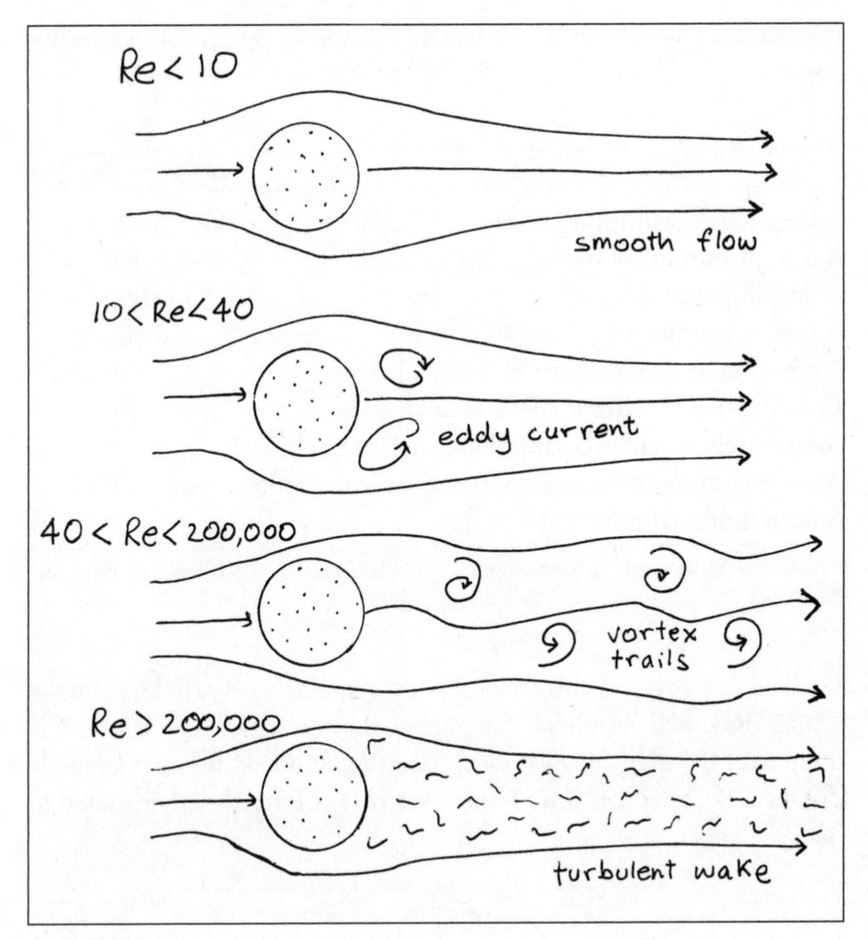

(Aatish Bhatia)

Let's say you want to simulate the effect of wind on a new wing design. You build a scale model in the lab that is one-tenth the size of the actual wing. But remember how the Reynolds number is defined. If you shrink the size of the wing by a factor of ten, you have to increase the wind speed by the same factor in order to keep the number fixed. The key point is that systems with the same Reynolds number have essentially the same nature of flow. If you didn't account for this, your wing would be quite a disaster.

How would a biologist use this idea? Well, nature presents us with organisms that cover an incredible range of sizes, from the tiniest microbes to blue whales. Here is a table of Reynolds numbers across this range.

	Reynolds Number
A large whale swimming at 10 m s^{-1}	300,000,000
A tuna swimming at the same speed	30,000,000
A duck flying at 20 m s^{-1}	300,000
A large dragonfly going 7 m s^{-1}	30,000
A copepod in a speed burst of 0.2 m s^{-1}	300
Flapping wings of the smallest flying insects	30
An invertebrate larva, 0.3 mm long, at 1 mm s^{-1}	0.3
A sea urchin sperm advancing the species at 0.2 mm s^{-1}	0.03
A bacterium, swimming at 0.01 mm s^{-1}	0.00001

(Adapted from Steven Vogel, *Life in Moving Fluids: The Physical Biology of Flow*, Princeton University Press, 1996)

The list covers fourteen orders of magnitude. A whale swims at a huge Reynolds number. This means that inertial forces completely dominate. If it flaps its tail once, it can coast ahead for an incredible distance. Bacteria live at the other extreme, as Purcell demonstrated in his 1977 paper.

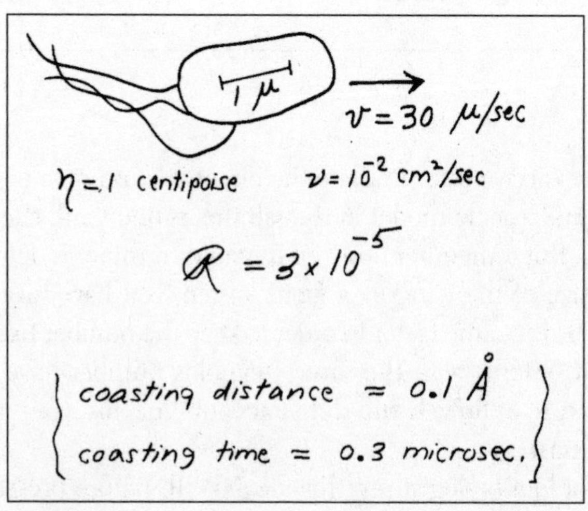

(Edward Purcell)

Eels and sperm may look similar, but their method of movement is very different, as their Reynolds numbers are far apart. So how would it *feel* to swim like a sperm, or a bacterium? To answer this, you have to get down to their Reynolds number. We can't change our size, but we can shrink our Reynolds number by swimming in a very viscous fluid. Purcell estimated that you would have to submerge yourself in a swimming pool full of molasses, and move your arms at the speed of the hands of a clock. (Don't try this at home. Swimming in molasses is not a good idea, as was sadly discovered in 1919 by the victims of Boston's Great Molasses Flood.) Under these conditions, if you managed to cover a few meters in a few weeks, then you qualify as a low Reynolds number swimmer.

This clearly isn't a hospitable environment for creatures of our size. Yet this is the scale of the task that microbes such as sperm face simply to get around.

(Edward Purcell)

In fact, it's even harder.

Here's a simple experiment. Place a few drops of dye, each a different color, in a large tube of glycerine. Then carefully stir the glycerine in one direction, say clockwise. What you get is a well-mixed concoction of dye and glycerine. The drops of colored dye have all merged together into a single multicolored band. Now comes the shocking part. If you next stir the glycerine in the opposite direction, something incredible happens—the dye and the glycerine begin to unmix. Eventually, you are left with the neatly arranged floating drops of dye that you began with.

The reason that the colors come back to where they start is because at low Reynolds number, flow is reversible. Because inertial forces are so small, certain terms drop out of the complicated fluid flow equations. The equations simplify considerably, and not only are they now solvable, they don't depend on time anymore. Everything flows smoothly, and any motion in one direction could just as easily have happened in reverse.

Our common-sense intuition is built around a medium Reynolds number world. Imagine filming the turbulent flow of smoke as it rises from a cigarette. If you were to play back this scene in reverse, things would look really strange. Spirals would uncurl and sinuous strands would weave together as the smoke winds back into the cigarette. Instinctively, we know this isn't right. But the low Reynolds number world is a simpler and more orderly place. Think of filming the experiment with the drops of dye and glycerine. When you play back the video, there's no way to tell whether you're watching it in reverse.

This reversibility has a surprising consequence. It means that anything that swims using a repeating flapping motion can't get anywhere. If it moves forward in one stroke, the other stroke will bring it right back to where it started. Scallops swim by opening their shells and snapping them shut. In low Reynolds number, the equivalent of a scallop can't get anywhere.

So how, then, do microbes manage to get anywhere? Well, many don't bother swimming at all; they just let the food drift to them. This is somewhat like a lazy cow that waits for the grass under its mouth to grow back. But many microbes do swim, and they make use of remarkable adaptations to get around in an environment that is entirely alien to us.

One trick they can use is to deform the shape of their paddle. By cleverly contorting the paddle to create more drag on the power stroke than on the recovery stroke, single-celled organisms like paramecia break the symmetry of their stroke and thus elude the scallop conundrum.

Navier-Stokes:

$$-\nabla p + \eta \nabla^2 \vec{v} = \rho \frac{\partial \vec{v}}{\partial t} + \rho(\vec{v} \cdot \nabla)\vec{v}$$

If $\mathcal{R} \ll 1$:

Time doesn't matter. The pattern of motion is the same, whether slow or fast, whether forward or backward in time.

The Scallop Theorem

(Edward Purcell)

Indeed, this is how the flapping structures known as cilia thrust a cell forward: they flex.

There is an even more ingenious solution that has been hit upon by bacteria, sperm, and other microbes. Rather than using cilia, which are essentially flexible paddles, these cells adopt a different strategy: they use a corkscrew for a propeller. Just as a corkscrew used on a wine bottle converts winding motion into motion along its axis, these organisms spin their helical tails (flagella) to push themselves forward.

But don't expect to see human swimmers doing "the corkscrew" anytime soon. This strategy works only at low Reynolds numbers, where water "feels" as thick as cork.

Motion in this viscous world is counterintuitive and puzzling. By applying science, we can imagine what it must feel like to be very small. And we can work out how to build tiny ships in such a world. But evolution has beaten us to the punch, and microorganisms have evolved intricate and wonderful structures that pulsate rhythmically and take advantage of the quirks of physics at this scale.

NOTES

1. Edward M. Purcell, "Life at Low Reynolds Number," *American Journal of Physics* 45, no. 1 (1977): 3–11.
2. Osborne Reynolds, "An Experimental Investigation of the Circumstances Which Determine Whether the Motion of Water Shall Be Direct or Sinuous, and of the Law of Resistance in Parallel Channels," *Philosophical Transactions of the Royal Society of London* 174 (1883): 935–82.
3. Kerstin Avila et al., "The Onset of Turbulence in Pipe Flow," *Science* 333, no. 6039 (2011): 192–96. doi:10.1126/science.1203223.

In addition to learning about this subject from the above papers, I learned a lot from a wonderfully engaging and thoroughly researched book by Steven Vogel: *Life in Moving Fluids: The Physical Biology of Flow*, 2nd ed. (Princeton, N.J.: Princeton University Press, 1996).

The theme of this post came from reading the following wonderful out-of-print book that I discovered in the basement of the Strand bookstore in New York City: *On Size and Life*, by Thomas A. McMahon and John Tyler Bonner (New York: Scientific American Books / W. H. Freeman, 1983).

AATISH BHATIA graduated from Swarthmore College in 2007 and has since been pursuing a Ph.D. at Rutgers University. He enjoys teaching and writing about science at www.empiricalzeal.com. This post brings together his passions for physics, biology, and white-water kayaking.

5

INCREDIBLE JOURNEY

LEE BILLINGS

Gliese 581d. HD 85512b. Kepler-22b. They have unfamiliar names, but these exoplanets could be relatively familiar places. They reside in orbits around their stars where liquid water could exist and life as we know it could thrive. No one knows for sure, but these worlds could be rather like Earth. What seems more certain by the day is that they are glimpses of a future in which, at long last, life is found elsewhere in the great beyond, and the cosmos suddenly becomes slightly less lonely.

If and when we locate another pale blue dot circling a nearby star, for many people the next logical step would be to attempt to send people or machines there for direct investigation. It sounds simple enough, to send a spacecraft from point A through mostly empty space to point B. The Moon hangs shining in the sky along with the stars, and we've already sent explorers there, as well as robotic emissaries to all the solar system's planets. Reaching the stars shouldn't be that much harder—but it is.

Consider the problem from the simple viewpoint of velocity. It's easy to forget that until very recently, the fastest anyone had ever traveled on planet Earth was almost certainly about 125 miles per hour (mph), the terminal velocity of a plummeting human form past which air resistance impedes further acceleration. But then our species learned to build machines that use the fossilized sunlight in coal, gas, and oil to go even faster.

In 1906, a Bostonian named Fred Marriott finally surpassed the millennia-old record—and lived to tell about it—traveling over 125 mph in a steam-powered car across the sands of Daytona Beach, Florida.

Scarcely forty years later, a West Virginian test pilot named Chuck Yeager flew a rocket-propelled plane at more than 650 mph, faster than the speed of sound at 42,000 feet. A decade after that, gargantuan rockets were accelerating men and machines to nearly 17,000 mph, fast enough to orbit Earth and gain a god's-eye view of the planet. That's how we sent astronauts to the Moon and robotic probes to other planets. Surely we can go even faster and undertake interstellar voyages.

But space is vast, and even the distance to the nearest star is mind-boggling. Let's say the Sun is the size of a large orange, four inches in diameter. Place the orange on the ground, walk eleven yards away, and you're in Earth's orbit. Finding our planet might prove challenging—it would be the size of a grain of sand: one millimeter, or four-hundredths (.04) of an inch. The walk out to Pluto, a speck of dust ten times smaller than our sand-grain Earth, would be nearly a third of a mile, and along the way you'd be lucky to encounter any of the planets. Even the largest, Jupiter, would be no bigger than a small marble.

From Pluto in this scale model, to reach the nearest star system, Alpha Centauri, you'd have to travel some 1,800 miles: roughly the distance between Memphis and San Francisco, or about how far you'd have to dig straight down into Earth before reaching its outer core. At this scale, light, the fastest thing in the universe, would travel through space at just over eight-tenths (0.8) of an inch per second. In actuality, light travels at 186,000 miles per second, and requires nearly four and a half years to reach Alpha Centauri from our solar system.

Today, the fastest humans on Earth and in history are three elderly Americans, all of whom Usain Bolt could demolish in a footrace. They're the astronauts of *Apollo 10*, who in 1969 re-entered Earth's atmosphere at a velocity of 24,792 mph upon their return from the Moon. At that speed you could get from New York to Los Angeles in less than six minutes.

Seven years after *Apollo 10*, we hurled a probe called *Helios* 2 into an orbit that sends it swinging blisteringly deep into the Sun's gravity well. At its point of closest approach, the probe travels at almost 160,000 mph—the fastest speed yet attained by a man-made object.

The fastest outgoing object, *Voyager 1*, launched the year after *Helios* 2. It's now almost 11 billion miles away, and travels another 11 miles farther away each and every second. If it were headed toward Alpha

Centauri (it's not), it wouldn't arrive for more than 70,000 years. Even then, it wouldn't be able to slow down. Of the nearest 500 stars scattered like sand around our own, most would require hundreds of thousands of years (or more) to reach with current technology.

Space and Ships

Part of the problem is rocketry. An inescapable fact of accelerating by venting material out of a nozzle is that it's not terribly efficient. Not even accounting for food, water, and other consumables, you must carry all your fuel along with you, and the faster you wish to go, the more fuel you'll need—fuel that itself requires additional amounts of fuel to accelerate the additional mass. We've already almost maxed out the velocities attainable through Apollo-style chemical rockets.

But even so, there are no insurmountable physical barriers preventing people and machinery from going much, much faster than the pioneers of forty and fifty years ago. A few idealistic scientists and engineers around the world still obsessively concoct new ways of harnessing more energy, of achieving more velocity, of going faster and farther than anyone has ever gone before. Maybe the stars are within reach.

We already know how to build speedier and more efficient rockets powered by electricity instead of chemicals, but they won't do much to get us to nearby stars. For that, only a handful of schemes could suffice. Some researchers suggest building rockets fueled by antimatter, an energy source so potent that the amount required to send you on a month-long crossing to Mars would be measured in grams. Others call for constructing gossamer-thin "sails" in space, each hundreds of miles wide, which would ride on powerful laser or particle beams out of our solar system. These options and their more exotic variations theoretically offer velocities that are a significant fraction of the speed of light.

Sadly, while the physics may be on our side, the economics aren't. Based on present production rates and costs, producing and storing enough antimatter to fuel an interstellar mission would quite possibly bankrupt the planet. As for an interstellar sail, such an endeavor would dwarf the largest single piece of space-based infrastructure yet built, the International Space Station, a construction project that has so far cost an estimated $150 billion. Constructing an interstellar sail would probably cost far more—and that's not including the truly astronomical electric bill associated with powering the multimillion-gigawatt laser that would need to shine on the outbound sail for years on end.

At present, the most economically viable fast boat out of the solar system would probably be a spacecraft propelled by regular pulses of detonating atomic explosives. We do, after all, already have plenty of nuclear bombs lying around for no other real purpose than destroying civilization. Perhaps it's not unreasonable to co-opt them for a more productive endeavor. The U.S. government actually funded a study of the concept in 1958, an ambitious program called Project Orion that seriously proposed, among other things, building a nuclear-pulse spacecraft that could send humans to the moons of Saturn as early as the 1970s. But legitimate concerns over radioactive fallout and the dual-use possibilities of miniaturized thermonuclear explosives forced Project Orion's eventual cancellation.

Flying High

A new effort to design a nuclear-pulse spacecraft began in September 2009: Project Icarus. Icarus is the offspring of an earlier highly regarded study, the 1970s-era Project Daedalus, named after a craftsman from Greek mythology who flew free from imprisonment on wings he constructed out of feathers and wax. Both projects plan spacecraft that would voyage to the stars propelled by thermonuclear fusion.

Fusion occurs when nuclei of light elements like hydrogen or helium are slammed together with such force that they merge, releasing a flood of energy. It's a process that creates and destroys. It's what gives hydrogen bombs their fearsome power, but it also is how stars shine, glomming together light nuclei in their cores to form heavier elements. Stellar fusion is what made the calcium in your bones, the carbon in your DNA, and the oxygen that you breathe.

If fusion reactions could somehow be used in a propulsion system, they could accelerate a spacecraft to perhaps 10 percent the speed of light. Daedalus envisioned replicating the pressure and heat of a star's core via arrays of high-powered lasers that would focus on small fuel pellets, compressing them past the fusion threshold and channeling the resulting plasma through a magnetic nozzle to produce thrust. The Icarus team is considering that approach, but has yet to decide on its thermonuclear propulsion method of choice.

Like Daedalus before it, Icarus is a project run entirely by volunteers, scientists and engineers who spend their idle time dreaming of starflight and performing laborious calculations to learn how it might be practically achieved. But unlike the Daedalus volunteers, who relied on the liberal use of slide rules, brand-new HP-35 calculators, and an occasional sketch on the back of a bar napkin, the Icarus team is leveraging the power of more than thirty additional years of technological progress.

Our advances in velocity may have petered out over the past few decades, but our prowess in information processing and communication has steadily accelerated.

Each individual volunteer on the Icarus team today could marshal more computing power than was available to most nation-states in the 1970s, and can electronically access the bulk of the world's accrued scientific and technical knowledge within seconds. Rather than gathering in pubs, the volunteers are formulating their starship design via Internet telephony, private messaging forums, and the occasional post on the official Icarus blog. Still, while anyone can crunch numbers, actually building a starship would consume a large chunk of Earth's entire economy, and likely would require creating massive economies off-world.

In the Daedalus plan, for instance, constructing a 650-foot-long, 4,000-ton spacecraft in Earth orbit was actually the easy part—never mind that such a ship would be roughly the same size as one of the U.K.'s *Queen Elizabeth*–class aircraft carriers. The harder task was acquiring 50,000 tons of the necessary thermonuclear fuel, an isotope of helium that is vanishingly rare on Earth. The Daedalus solution was to harvest the fuel from gas-giant planets such as Jupiter, by building and operating a fleet of balloon-borne robotic extraction factories in their atmospheres. In other words, the easiest way the Daedalus volunteers found to fuel their starship was, in effect, the industrialization of the outer solar system.

Additional obstacles abound. Traveling at a significant fraction of light-speed can be compared to staring down the barrel of a gun. Running into a small piece of dust, or, heaven forbid, a sand grain, could cause catastrophic damage. The preferred Daedalus countermeasure was a fifty-ton beryllium shield placed at the ship's prow. Even if no damaging impacts occurred, a starship on a mission of decades or centuries would still require maintenance as parts and components wore out or broke down. For Daedalus, the solution was to pack a number of autonomous robotic wardens onboard the spacecraft to repair damage as it occurred. Creating such artificially intelligent robots capable of tending a starship for decades on end might be a bit more difficult than designing a Roomba to autonomously vacuum your living room.

And all that effort would send only a 500-ton payload, sans humans, strictly on a one-way flyby of a star. There would be no slowing down, stopping, or returning home. The Daedalus probe would fly through the alien star system in a matter of hours, gradually trickling data homeward via a parabolic radio antenna. After absorbing untold treasure, time, and talent to reach another star, the Daedalus starship would have sent back scarcely more than the cosmic equivalent of a postcard.

Icarus has upped the ante: the team intends to design a starship that can enter into orbit around its target star, perhaps to monitor any potentially habitable planets there, and then, somehow, send large amounts of data back to Earth. Suffice it to say, engineering at these scales makes "rocket science" look like child's play.

Consider the disruptive, unanticipated effects of the technologies a project like Icarus currently uses—not even including the ones it hopes to eventually employ for its starship. The ubiquitous computing and information networking that now allows Icarus to break out of local pubs and stretch across the world also seems to be turning many people's focus inward, simultaneously connecting and unweaving the world.

The velocity of our technology may ultimately be too fast, rather than too slow, and like the Daedalus probe of yore, our civilization could accelerate past its passion for frontiers in a flash, never to return. We could all too easily become lost in the virtual worlds we make for ourselves, and lose interest in the stars. Or, more probably, through gradual decay or sudden disaster we could squander our resources and experience profound and irreversible technological regression. Sometimes I pessimistically hold with some combination of these two extremes.

Given the magnitude and number of extreme technological and economic challenges that must be overcome to achieve starflight, it's difficult to imagine what, in fact, a civilization capable of interstellar travel would look like. Probably not much like us—more than anything else, projects like Icarus and Daedalus seem to tell us that we are presently as distant from interstellar travel as the stars are from Earth. And, at least until our culture's prioritization of short-term profit once again aligns with pushing the limits of the ultimately possible, that's likely to remain the case.

Perhaps someday one of these familiar starship designs will take us out of the solar system on voyages to other living planets, other cosmic oases, strewn among the stars. Or maybe all the methods conceived today will bear no more resemblance to actual starships than airplanes have to birds. Either way, it's worth remembering that the 100,000-year duration of interstellar voyages we could undertake right now is but the blink of an eye in cosmic terms. It may be more effective to adapt our expectations to those timescales, to master long-term planning rather than brute-forcing our way to Alpha Centauri.

In expanding outward into space, patience, not velocity, may be the greatest virtue. After all, we're already on an interstellar spacecraft called Earth, sailing with the Sun and its retinue of other planets around the Milky Way in circuits lasting 250 million years. Only by carefully preserving and cultivating the relatively bountiful and accessible resources of our planet and the solar system will we ever escape their confines. For now, it's wise to reflect that in our headlong rush to go ever faster and farther, we may be only fooling ourselves.

LEE BILLINGS is a freelance science journalist and the author of a forthcoming book on the search for Earth-like planets elsewhere in the cosmos. He lives in New York.

6

IN THE SHADOWS OF GREATNESS

BIOCHEM BELLE

Even if you're not a chemist, you might have heard of Fritz Haber in a general chemistry course. A German chemist in the early twentieth century, Haber is most widely known for the reaction that bears his name, the Haber or Haber-Bosch process, the first method for synthesizing ammonia from nitrogen and hydrogen gases.

It may seem trivial now, but at the time, chemists had been trying to do just that for over a century. Haber and collaborator Robert Le Rossignol found that under high pressure and high temperature, the gases would react to form ammonia. The formation of ammonia was still quite slow, but the pair subsequently discovered that the rare metal osmium accelerated the reaction. Following a successful demonstration of the process, Carl Bosch and Alwin Mittasch, working for Germany's largest chemical company, BASF, were tasked with industrializing the process. Osmium was expensive and its supply limited. Mittasch tried 4,000 catalysts and found that a mixture of iron and metal oxides, a much cheaper and more abundant alternative, could be used.

The first industrial unit started production in September 1913, generating up to five tons of ammonia every day. At this scale, ammonia could be readily converted to nitrates, which were (and still are) used in fertilizers and explosives. Prior to industrial production of ammonia, Germany's primary source of nitrates was saltpeter mines in Chile, a supply expected to be depleted within thirty years. Moreover, supply from the mines was blocked by the British Royal Navy after the start of World War I in 1914.

An oft-quoted claim is that at the start of World War I, Germany's stockpile of saltpeter could only have supplied munitions for just over a year, and that without the Haber process, Germany would have been forced to end the war in 1916. Jerome Alexander, a chemist of the day, even suggested that "without the Haber process it is doubtful if Germany would have started the war, for which she carefully prepared."[1]

Haber's impact on the war front did not stop there. Haber was a fierce patriot, devoting his time, intellect, and force to Germany's war effort, and he was appointed to the War Department for Raw Materials as head of chemistry. By numerous accounts, he lived by his credo, "In peace for mankind, in war for the fatherland!" Realizing that tear gas would be ineffectual on the frontlines, Haber recommended using chlorine gas to flush enemy soldiers out of the trenches. He committed himself to turning this idea into reality, even though it was in direct contravention of the Hague Conventions. He oversaw every aspect of the effort: manufacturing, testing, installation.

The first test on the frontlines came at the second Battle of Ypres, ushering in a new era of warfare. Over 150 tons of chlorine gas were released within ten minutes in each of two deployments in two days. The result: 15,000 casualties with 5,000 dead, including many German soldiers, because gas masks did not reach the front prior to the attack.

Haber continued to develop chemical warfare agents and tactics. He was not even dissuaded by the suicide of his wife, who considered chemical warfare barbaric (although it's unclear how much Haber's work contributed to her suicide). It seems Haber hoped that chemical warfare would bring a swift victory, and thereby an early end to the war. Instead, by the end of the war, both sides were using chemical weapons, contributing to more than 90,000 deaths and over 1 million casualties. Yet even after the war, Haber continued to oversee development and production of chemical agents.

Haber's war contributions were at the center of the consternation surrounding the 1918 Nobel Prize in Chemistry (announced in 1919). Here stood a man who months before had been labeled a war criminal (though the charges were dropped), now being lauded for his scientific prowess. The prize was awarded for "the synthesis of ammonia from its elements," and the committee highlighted the importance of this accomplishment to agriculture. There was no mention of the use of ammonia for making explosives or of Haber's campaign for chemical warfare.

This brings us to the question: What makes scientists great? Do we consider primarily their greatest scientific contributions? The brilliance and creativity that brought them to the answer for a long-held question? A solution for a problem plaguing the field? Do we focus solely on how their science benefited society?

The staggering rise in world population over the second half of the twentieth century was supported in part by the Haber-Bosch process. By one estimate, in 2008, almost half the world's population was sustained by agriculture dependent on fertilizer made from Haber's ammonia. Saving countless individuals from starvation seems a noble achievement.

Yet, at times, there are alternative uses for scientific discoveries, and unforeseen consequences. From my readings, the motivation behind Haber's desire to make ammonia from air remains unclear. Some imply that he pursued this work with the knowledge—and maybe even the intention—that it would be used for war. Some are more generous, suggesting he did it to help mankind, and others assimilated the work for more nefarious purposes. Max Perutz wrote:

> By a terrible irony of fate, it was his apparently most beneficent invention, the synthesis of ammonia, which has also harmed the world immeasurably. Without it, Germany would have run out of explosives once its long-planned blitzkrieg against France failed. The war would have come to an early end and millions of young men would not have been slaughtered. In these circumstances, Lenin might never have got to Russia, Hitler might not have come to power, the Holocaust might not have happened, and European civilization from Gibraltar to the Urals might have been spared.[2]

So do we ignore the dark side of science and its practitioners? How should our perception of the morality of scientists' actions affect their standing as the greats of science? Setting aside Haber's motivation for making ammonia, he had a very good idea of what he was promoting when he began developing toxic gases and delivery methods for them.

And what of the brilliant minds involved in the making of the atomic bomb? Do we more easily overlook the "sins" of scientists if, later in life, they express remorse over their contributions to death and destruction? Do we excuse culpability because we consider a cause just? Is it the level of involvement that affects our perspective—whether a scientist was active, complicit, or simply stood by and did nothing? Should we consider the inevitability of a discovery? Is a discovery less great—or the blame for negative impact less dire—if others were nipping at the discoverer's heels? Perutz posited:

> Haber's synthesis of ammonia for fertilizer was an extremely important discovery, but, unlike relativity, it did not take a scientist of unique genius to conceive it; any number of talented chemists could, and no doubt would, have done the same work before very long.[3]

I suspect the same would hold true for chemical warfare. If it hadn't been Haber, it would have been someone else. And Allied forces responded with similar tactics in turn. Yet it is Haber who is known, to this day, as the father of chemical warfare.

Moving beyond the questionable and destructive work of great scientists, what of the more eccentric ideas or less rigorous investigations of famous scientists? How do these alter our concept of greatness? Haber was but one of many who spent years trying to extract gold from seawater. Or consider Linus Pauling's promotion of vitamin C as a health remedy. Perutz noted, "It seems tragic that this should have become one of Pauling's major preoccupations for the last 25 years of his life and spoilt his great reputation as a chemist."[4] Yet Perutz still considered Pauling the greatest chemist of the twentieth century, in spite of this failing.

Some may feel that great scientists should not have such weaknesses, but I find it easier to pass over these more innocuous shortcomings. In my mind, such idiosyncrasies remind us that these legends were indeed human and, thus, flawed. They also provide cautionary tales of how even the brightest minds can fall into the trap of searching for the answer they want to find, rather than the answer the data yield.

These questions are hardly new. But as we consider who the greatest individuals in our field might be, we would do well to contemplate not only their magnificent achievements but also the things hidden in their long shadows.

NOTES

1. Jerome Alexander, "Nobel Award to Haber. Source of the Resentment Felt in Allied Countries," letter to the editor, *New York Times*, February 3, 1920.
2. Perutz, "Friend or Foe of Mankind?" in *I Wish I'd Made You Angry Earlier*, 16.
3. Ibid.
4. Perutz, "What Holds Molecules Together?" in *I Wish I'd Made You Angry Earlier*, 187.

REFERENCES

Eckstrand, Å. G. "The Nobel Prize in Chemistry 1918." In *Nobel Lectures: Chemistry 1901–1921*. Amsterdam: Elsevier Publishing Company, 1966. www.nobelprize.org/nobel_prizes/chemistry/laureates/1918/press.html.

Erisman, Jan Willem, Mark A. Sutton, James Galloway, Zbigniew Klimont, and Wilfried Winiwarter, "How a Century of Ammonia Synthesis Changed the World." *Nature Geoscience* 1 (2008): 636–39. doi:10.1038/ngeo325.

Friedrich, Bretislav. "Fritz Haber (1868–1934)." Published in part in *Angewandte Chemie* (International Edition) 44, no. 26 (2005): 3957–61, and 45, no. 25 (2006): 4053–55. www.fhi-berlin.mpg.de/history/Friedrich_HaberArticle.pdf.

Gibson, Adelno. "Chemical Warfare as Developed During the World War and Probable Future Development." An Address at the Annual Graduate Fortnight of the New York Academy of Medicine, October 22, 1936. *Bulletin of the New York Academy of Medicine* 13, no. 7 (1937): 397–421.

Perutz, Max F. "Friend or Foe of Mankind?" In *I Wish I'd Made You Angry Earlier: Essays on Science, Scientists, and Humanity*, expanded ed., 3–16. Cold Spring Harbor, N.Y.: Cold Spring Harbor Laboratory Press, 2003.

——. "What Holds Molecules Together?" In *I Wish I'd Made You Angry Earlier*, 179–88.

By day, BIOCHEM BELLE, postdoc, (usually) revels in biochemical research. She muses on science and life therein at http://biochembelle.wordpress.com.

7

A VIEW TO A KILL IN THE MORNING

DEBORAH BLUM

In 1937, inspired by a tragic accident, a New York pathologist came up with the scenario for a perfect murder.

His idea was based on the deaths of five longshoremen, their bodies found in the cargo hold of a steamer docked on the East River. The boat had been carrying cherries from Michigan. The men had been bunking in the room where the fruit was stored, and to the shock of their coworkers as they started to unload the cherries, all five were found lifeless in their beds.

When investigators from the New York City medical examiner's office arrived, they discovered that the fruit had been chilled by placing large containers of "dry ice" in the storage area. Dry ice is, of course, not ice at all but carbon dioxide (CO_2) in its solid form, resembling breathtakingly cold chunks of frosted glass. At standard atmospheric pressure, water (H_2O) freezes to ice as temperatures slip just below 32°F. Carbon dioxide solidifies (a process called deposition) at –109.3°F.

As it warms—say, as it sits in a fruit storage area—it begins returning to its gaseous state, a transition known as a phase change. The solid chunks shrink without the seeping wet of melting water ice, hence the name "dry ice" (patented in 1924 by the Dry Ice Corporation of America). Instead, there is a steady seep of gas into the surrounding air. Mostly this is nothing to worry about—unlike its chemical cousin, carbon monoxide (CO), carbon dioxide is not acutely poisonous—and, in fact, the chilly vapors lifting off dry ice have been used to create a fog effect in places ranging from theaters to Halloween parties.

But there is a risk. Carbon dioxide is denser than oxygen-rich air and can, notably in confined spaces, essentially displace the breathable atmosphere, settling into its surroundings like an invisible but suffocating blanket.

And this is what the city's medical examiners realized had happened to their dead longshoremen. The men's blood was "saturated with carbon dioxide and the men had obviously died from asphyxia," explained deputy medical examiner M. Edward Marten in his 1937 book, *The Doctor Looks at Murder.*

Still, he added, finding CO_2 in the blood went only partway to solving the puzzle. Carbon dioxide is always naturally present in human blood. It's a by-product of the way we use oxygen to metabolize sugars, and as CO_2 builds up, we exhale it away. If a person is murdered by suffocation and cannot exhale, the gas also builds up in the blood:

"Exactly the same autopsy picture would have been found if the men had died from being smothered by holding, say, a pillow over their mouths.

"This brings up a rather interesting possibility for a method of murder that would be extremely difficult to detect," Marten continued. "I pass this on, for what it is worth, to writers of detective stories."

In his scenario, a sleeping or heavily intoxicated person slumbers in bed. The killer places a bucket packed with dry ice on the floor, shuts all windows, and closes the door tightly as he leaves. Within a few hours, the victim suffocates. When someone else opens the door, normal air refills the room, whisking away all trace of the murder weapon: "The trick is that when dry ice evaporates it leaves absolutely no trace behind, so that the investigating detectives would find nothing except a dry and completely empty pail."

Still, Marten considered that a better tip for fiction writers than for real-life killers. The purchase of dry ice was easy to track, the material so cold as to bring on frostbite if handled improperly, and an ideally airtight room almost impossible to find. And someone, after all, might wonder about that peculiarly placed empty pail.

Nevertheless, I'd like to take this moment to pay tribute to carbon dioxide, as one of the most important—and dangerous—gases on the planet. We tend to discount its lethal potential by contrast with its toxic chemical cousin, carbon monoxide (CO).

Thanks to its ability to block the hemoglobin in red blood cells from carrying and circulating oxygen in the bloodstream, carbon monoxide drifting from faulty heaters, generators, cars accidentally left running, furnaces, and other fuel-burning machinery is estimated to kill some 500 people in the United States every year and to send thousands of others to doctors and hospitals. Carbon dioxide is not as direct in its deadly effects.

And we tend to discount carbon dioxide as an actual poison because we're focused instead on all the other ways it can—and does—cause trouble. These days, it's best known as a "greenhouse gas," for its ability to trap heat in the atmosphere.

Numerous studies have found that levels of CO_2 have risen steadily due to human activities—ranging from industrial burning of carbon-rich fuels to deforestation to agricultural practices. The U.S. Environmental Protection Agency estimates , for instance, that in 2005, "global atmospheric concentrations of CO_2 were 35 percent higher than they were before the Industrial Revolution." Although other gases are also linked to the current scenario of human-induced climate change, carbon dioxide is considered by many to be the most important factor.

Of course, even that doesn't really give full credit to the ways that carbon dioxide can alter the environment. For instance, scientists calculate that our oceans absorb a good proportion of the gas generated by human activities. Unfortunately, when you dissolve CO_2 into H_2O you rather logically end up with the compound H_2CO_3, better known as carbonic acid. If you don't recognize it, it's the rather weak acid found in carbonated soft drinks (although not so weak that countless middle school students haven't studied its corrosive effect on everything from teeth to lug nuts).

It's not surprising, then, that marine biologists have expressed alarm about the increasing acidification of the oceans. One recent report I found evaluated the effects of predicted increases in carbonic acid levels on California mussels, finding a notable thinning of their shells and a decrease in their overall size. A study conducted by Norwegian scientists also found that mussel larvae decreased in size, but the study suggested optimistically that the effects might be mitigated if, as seemed probable, only the larger larvae would survive. Of course, mussels are far from the only species at risk, as a National Science Foundation report concluded recently, listing everything from coral to marine algae.

Which is just another way of saying that we're deep into a global chemistry experiment with one of nature's most important, troublesome, and occasionally lethal chemical compounds.

There are many reasons, in fact, why we should regard carbon dioxide with respect, if not downright wariness.

And, certainly, one finds that kind of response at a subconscious level. A fascinating experiment a couple of years ago found that just inhaling a small amount of carbon dioxide triggered a fear response in mice. And there's an equally fascinating wealth of research about the relationship between human panic disorders and CO_2 inhalation. Far beyond my New York murder scenario, there's rather horrifying evidence that occasionally this can be a panic-worthy gas.

The best example of that comes from a real-life event, a catastrophic natural release of carbon dioxide in Lake Nyos in Cameroon during the summer of 1986. Beneath that beautiful lake, geothermal seeps release CO_2 into the deep lake waters, normally trapped near the bottom by pressure and cold. But in this case, apparently, the lake became oversaturated with the gas, and on August 21, 1986, the lake waters effectively turned over, carbon dioxide fizzing explosively upward, the waters of Lake Nyos turning a startling red as iron deposits were stirred about.

The rapidly released gas settled in a suffocating layer into the valleys around the lake. So many people died—an estimated total of 1,746—that eventually the website Snopes.com felt compelled to investigate. Snopes reported that the event was real and, in fact, not the only case of suffocation deaths due to carbon dioxide seeps at lakes in Cameroon. Since that time, in fact, measures have been taken to maneuver the gas out of the lakes.

And this, of course, brings me back to plotting a CO_2 murder.

I discovered Martens's theory and his long-out-of-print book while researching early twentieth-century forensic toxicology a few years ago. At first, I just liked the improbability, a medical examiner cooking up a supposedly unsolvable murder. But what's stayed with me is the implicit message—that we're talking about a dangerous compound.

It's a message to remember as we move deeper into our global experiment in greatly increasing the amount of carbon dioxide in our environment. For those who haven't taken the experiment seriously—in my opinion, still far too many—it's a reminder that they should start doing so now. And for those who need no reminder of the bigger carbon dioxide picture, I think I can still pass along at least one useful tip: in case of strangers bearing buckets of dry ice, sleep with your windows open.

DEBORAH BLUM is a Pulitzer Prize–winning science writer and the author of five books, most recently *The Poisoner's Handbook: Murder and the Birth of Forensic Medicine in Jazz Age New York*. A professor of journalism at the University of Wisconsin–Madison, she also writes for a range of publications, including *Slate*, *Time*, *Scientific American*, *The Wall Street Journal*, and the *Los Angeles Times*. She blogs about chemistry and culture at *Elemental* at Wired.com.

8

IT'S SEDIMENTARY, MY DEAR WATSON

DAVID BRESSAN

On February 20, 1949, Mrs. Henrietta Helen Olivia Roberts Durand-Deacon, a sixty-nine-year-old wealthy widow, disappeared from the Onslow Court Hotel located in South Kensington, London. The police interviewed the residents, and soon forty-year-old John George Haigh became a suspect, as he was the last person to have seen the woman alive and was known already to the police for crimes of fraud and thievery. He led the police to an old storeroom on Leopold Road in Sussex, where they discovered strange and suspicious tools: a revolver, some rubber protective clothing, and three containers filled with sulfuric acid.

Soon afterward, during an interrogation, Haigh suddenly confessed to an incredible crime: "Mrs. Durand-Deacon no longer exists. She has disappeared completely, and no trace of her can ever be found again. I have destroyed her with acid. You will find the sludge which remains on Leopold Road. But you can't prove murder without a body."

Fortunately Haigh, in his euphoria, ignored one important fact: the law doesn't require a corpse to incriminate a suspect; it requires a corpus delicti—evidence that a murder happened. The eminent forensic pathologist Keith Simpson examined carefully the ground at the supposed crime scene. He noted something unusual—a small pebble that he described as follows: "It was about the size of a cherry, and looked very much like the other stones, except it had polished facets."

This pebble was unlike other rocks found at the site. Soon Simpson realized that he had found the evidence to prove the murder. The pebble was a gallstone from poor Mrs. Durand-Deacon. Gallstones can form from calcium salts and organic substances in the gallbladder; the organic sludge that covers them protected the pebbles from being dissolved in the acid.

John George Haigh, who was ultimately suspected of committing an entire series of murders, was sentenced to death by hanging.

This forensic case was an unusual example of how a pebble can help solve a crime. However, already in the mid–nineteenth century people realized that rocks, soils, and the science of geology could be used to reconstruct a crime, and could provide circumstantial evidence to connect a suspect with the crime scene. An 1856 issue of *Scientific American* reported the "Curious Use of the Microscope" to help clarify a case of thievery:

> Recently, on one of the Prussian railroads, a barrel which should have contained silver coin, was found, on arrival at its destination, to have been emptied of its precious contents, and refilled with sand. On Professor Ehrenberg, of Berlin [1795–1896, famous zoologist and geologist from Leipzig], being consulted on the subject, he sent for samples of sand from all the stations along the different lines of railway that the specie had passed, and by means of his microscope, identified the station from which the interpolated sand must have been taken. The station once fixed upon, it was not difficult to hit upon the culprit in the small number of employees on duty there.

Influenced by the rapid development of science, the British author Sir Arthur Conan Doyle introduced in 1887 a new kind of detective, who based his crime-solving abilities on the scientific and forensic clues that everybody acquired or left behind by touching objects, or simply by walking on muddy ground:

> Knowledge of Geology.—Practical, but limited. Tells at a glance different soils from each other. After walks has shown me splashes upon his trousers, and told me by their colour and consistence in what part of London he had received them.
>
> —Dr. Watson's description of Holmes's abilities in A *Study in Scarlet*

At about the same time as Sir Arthur Conan Doyle was publishing Holmes's fictional adventures, the Austrian professor of criminology Hans Gross (1847–1915) published various textbooks dealing with forensic investigation methods, trying to introduce a ubiquitous standard and consistent investigative approaches.

In his *Handbuch für Untersuchungsrichter als System der Kriminalistik* (*Handbook for Examining Magistrates as a System of Criminal Investigation*), published in 1893, he proposed that the police should carefully study geographical and geomorphological maps to infer possible sites where criminals could commit crimes or hide bodies—such as forests, ponds, streams, or sites with a well. Gross explained how the petrographic composition of dirt found on shoes could indicate where a suspect had previously been.

Based on these ideas, in 1929 the French physician and pioneering criminalist Edmund Locard (1877–1966) stated his basic "exchange principle" of environmental profiling—by which organic or inorganic substances found in the environment can connect a suspect with a crime:

> Whenever two objects come into contact, there is always a transfer of material. The methods of detection may not be sensitive enough to demonstrate this, or the decay rate may be so rapid that all evidence of transfer has vanished after a given time. Nonetheless, the transfer has taken place.

The German chemist Georg Popp (1867–1928) was the first investigator to solve a murder case by adopting the principles of Gross and Locard and considering soil as reliable evidence. In the spring of 1908, Margarethe Filbert was murdered near Rockenhausen in Bavaria.

The local attorney had read Hans Gross's handbook and knew Popp from an earlier case, in which Popp had connected a strangled woman to the suspect by mineral grains of coal and hornblende found in the mucus of the victim's nose and under the fingernails of the suspect. In the Filbert case, a local factory worker named Andreas Schlicher was suspected; however, he claimed that on the day of the murder he was working in the fields.

Popp reconstructed the movements of the suspect by analyzing the dirt found on his shoes. The innermost layer, thus the oldest, contained goose droppings and earth from the courtyard of the suspect's home. A second layer contained red sandstone fragments and other particles of a soil found where the body of the victim was discovered. The outer layer contained brick fragments, coal dust, cement, and a whole series of other materials also found on the site where the suspect's gun and clothing had been found. However, there were no mineral grains—fragments of porphyry, quartz, or mica—on the shoes. Since these were found in the soils of the field where Schlicher claimed to have worked the very same day, he was obviously lying.

In the last two decades, the significance of forensic geology has increased steadily. It is applied to connect single suspects to criminal cases, but also to trace the origin of explosives, drugs, or smuggled goods, including wildlife—not to mention the possible applications to detect violators of environmental law. Forensic geology has also proved valuable for reconstructing and uncovering modern war crimes.

In 1997 the United Nations International Criminal Tribune for the Former Yugoslavia (UN ICTY) began exhuming mass graves in northeastern Bosnia associated with the massacre of civilians in and around the town of Srebrenica in July 1995. Intelligence reports showed that three months after the initial executions of civilians, the primary mass graves had been exhumed and the bodies transported over a one- to three-day period to an unknown number of secondary grave sites.

To prosecute the suspects, it was necessary to prove that the now-recovered bodies came without doubt from Srebrenica, and that therefore the later dislocation of the graves was intended to hide these war crimes. Two grave sites were intensively studied, with samples of the grave fills and surrounding soils and bedrock collected. Soil samples can be screened by their content of minerals and rocks, the size and form of single mineral or rock grains, biochemistry of organic substances, microbiology, remains of invertebrates and plants, and pollen and spores preserved in the soil.

These various parameters can vary in so many ways that every soil can be regarded as unique. Comparing the parameters between samples recovered from the victim or the suspect and collected at the crime sites, it is possible to establish a unique connection between them.

During the investigations in Bosnia, a clast of serpentinite found in one of the secondary grave sites proved to be decisive evidence. This greenish rock connected one secondary grave site with only one primary site—the only place where an outcrop with a serpentinite dike could be found. Similarly, the presence or absence of particular clay minerals, depending on the surrounding geology of the primary burial site, confirmed or ruled out connections from the primary to the secondary sites.

These are only some examples of the application of forensic geology. The possible list of fascinating or strange cases would surprise even Sherlock Holmes himself. The only reasonable deduction: whenever a crime is committed, there could well be a minuscule mineral grain or a pebble that might one day give testimony against the criminal.

REFERENCES

Alden, Andrew. "Sherlock Holmes, Forensic Geologist." About.com Geology. http://geology.about.com/od/bookreviews/a/sherlock.htm.

Brown, Anthony G. "The Use of Forensic Botany and Geology in War Crimes Investigations in NE Bosnia." *Forensic Science International* 163, no. 3 (2006): 204–10.

Murray, Raymond C. "Collecting Crime Evidence from Earth." *Geotimes* 50, no. 1 (2008): 18–23. www.forensicgeology.net/science.htm.

Pye, Kenneth. "Forensic Geology." In *Encyclopedia of Geology*, edited by Richard C. Selley, L. Robin M. Cocks, and Ian R. Plimer, 261–73. Amsterdam: Elsevier, 2004.

Rogers, J. David. Forensic Geology Case Histories. Excerpted from Raymond C. Murray and John C. F. Tedrow, *Forensic Geology: Earth Sciences and Criminal Investigation*, revised ed. Englewood Cliffs, N.J.: Prentice-Hall, 1992. http://web.mst.edu/~rogersda/forensic_geology/Geoforensics%20Case%20Histories.htm.

Ruffell, Alastair, and Jennifer McKinley. "Forensic Geoscience: Applications of Geology, Geomorphology and Geophysics to Criminal Investigations." *Earth-Science Reviews* 69, no. 3–4 (2005): 235–47.

——. *Geoforensics*. Hoboken, N.J.: John Wiley & Sons, 2008.

Wagner, E. J. *The Science of Sherlock Holmes: From Baskerville Hall to the Valley of Fear, the Real Forensics Behind the Great Detective's Greatest Cases*. Hoboken, N.J.: John Wiley & Sons, 2006.

DAVID BRESSAN is a freelance geologist; his graduation project on rock glaciers in the Alps left him with a special interest in modern glacial environments and the development of geological concepts by geologists through historic times. Appropriately, he blogs on these topics in the *Scientific American* network blog *History of Geology*: http://blogs.scientificamerican.com/history-of-geology/.

9
MENSTRUATION IS JUST BLOOD AND TISSUE YOU ENDED UP NOT USING

KATE CLANCY

Dr. Béla Schick was a very popular doctor in the 1920s and received flowers from his patients all the time. One day he received one of his usual bouquets from a patient. The way the story goes, he asked one of his nurses to put the bouquet in some water. The nurse politely declined. Dr. Schick asked the nurse again, and again she refused to handle the flowers. When Dr. Schick asked his nurse why she would not put the flowers in water, she explained that she had her period. When he asked why that mattered, she confessed that when she menstruated, she made flowers wilt at her touch.

So, rather than consider the possibility that the nurse was offended that her skills and expertise were being wasted putting someone else's flowers in water, Dr. Schick decided to run a test. Gently place flowers in water, on the one hand . . . and have a menstruating woman roughly handle another bunch in order to really get her dirty hands on them.

The flowers that were not handled thrived, while the flowers that were handled by a menstruating woman wilted.

This was the beginning of the study of the menstrual toxin, or menotoxin, a substance secreted in the sweat of menstruating women.

This story begins far before Dr. Béla Schick and his menstruating nurse. Because the kind of bias that produces a doctor who can believe that menstrual toxins exist and launch a field of study of them based on some wilted flowers (if the story really did happen the way he tells it) did not come from one man alone. The cultural conditioning that has produced the idea that women are dirty, particularly during menses, is quite old. The Old Testament of the Bible claims that women are unclean when they menstruate, and menstrual huts exist in some cultures to separate out menstruating women from the rest of their group.

But some mark the beginning of our misunderstandings of female physiology in European-derived cultures with one book in particular, written in the thirteenth century: *De Secretis Mulierum*, "Of the Secrets of Women." This book was written by a man who claimed to be the monk Albertus Magnus but was most likely an impersonator (which is why most call the author of *De Secretis Mulierum* pseudo–Albertus Magnus, or pseudo-Albert).

So here are some winning quotes from this book, which was considered a premier text for several centuries, even though it is likely pseudo–Albertus Magnus never treated women and based much of his work on having dissected a female pig:

- "Woman is not human, but a monster."
- Menstruating women, especially older women, give off harmful fumes that will "poison the eyes of children lying in their cradles by their glance."
- Children conceived by menstruating women "tend to have epilepsy and leprosy because menstrual matter is extremely venomous."

De Secretis Mulierum went through at least eighty editions over several centuries (Lemay 1992). While it was not a strictly medical text, it is clear that it was both popular and influential. Do doctors refer to *De Secretis Mulierum* today? Of course not. But this book, to me, represents a broader cultural understanding that menstruation is dirty, that women are powerful, mysterious, dangerous, and subhuman.

So back to those menotoxins. Dr. Schick decided there was something nasty in the sweat of menstruating women. Others took up the cause. Soon people were injecting menstrual blood into rodents, and those rodents were dying (Pickles 1979). Others were growing plants in venous blood from menstruating women to determine phytotoxicity; the sooner the plants died, the higher the quantity of menotoxin assumed in the sample.

What's worse, the presence of the menotoxin in the female body began to expand beyond menstruation. Any woman who was post-menarcheal and pre-menopausal could be found to have the menotoxin in her system. She could not escape it: some reported that the menotoxin could be found in a woman's menstrual blood, but also in venous blood, sweat, and breast milk. One case study reports that a mother gave her child asthma because she was menotoxic during pregnancy (Perlstein and Matheson 1936), and several contended that colic was caused by menotoxin in breast milk (Montagu 1940; Perlstein and Matheson 1936).

Not only did the idea of the menotoxin become a ubiquitous menace around any reproductive-age woman, it began to explain pathology. So the menotoxin, which first was an explanation for the presence of menstruation in women, became a way of diagnosing women as ill . . . and again, since now all reproductive-age women could secrete it from any bodily fluid at any time, the state of being female essentially made one pathological.

Soon the idea began to take hold that the menotoxin indicated specific illnesses: "Dr. Schick and I discussed the possibility that the adult female diabetic out of control, the depressed adult female psychotic, and the adult female in the premenstrual phase secreted some common substance in their sweat," writes Helen Evans Reid (1974). Here you see premenstrual women compared directly with two pathological conditions: diabetes and psychosis. And all of these relationships, between menstruation and colic, asthma, wilted flowers, are largely based on observation, case reports, or poorly controlled experiments. When studies do not support the idea of the menotoxin, as with Freeman, Looney, and Small (1934) and two studies cited by Montagu (1940) that were not in English, they get dismissed as outliers (even though in Labhardt's case from Ashley Montagu, the sweat of men was often as toxic as that of menstruating women).

Now, I love science, and I love the scientific method. I think that the scientific method is one of the most useful ways of knowing out there. I have devoted my life not only to the study of the science of human evolution and female reproductive physiology but to increasing science appreciation and literacy in the general public. But here's the thing: science can be biased by the cultural conditioning of those who perform it, and those who tell it. The people who studied the menotoxin really, really wanted to believe in it, to the point that they would ignore negative results and overstate the power of their anecdotes and case studies. The study of the menotoxin spans at least sixty years—maybe ninety, depending on which references you consider legitimate—during which it has been debated in *Lancet* letters to the editor and published in several medical journals. But it's wrong, and looking at the evidence cited when it was popular, these scientists should have known that.

I wish I could say that the idea of the menotoxin was dead. But several contemporary hypotheses about the evolution of menstruation still in some way reflect the thinking that menstruation, if not women, is dirty and serves the purpose of expelling toxicity. Clarke (1994) proposed menstruation as a mechanism to expel unwanted embryos. Margie Profet (1993) argued that menstruation helped to expel sperm-borne pathogens, which made men the dirty party. This is why it's important to recognize that many ideas that seem intuitive to us at first derive from cultural conditioning and bias. (My favorite book on the topic is Emily Martin's *The Woman in the Body* [1980].)

Thankfully, the most accepted idea is that menstruation did not evolve at all but is a by-product of the evolution of terminal differentiation of endometrial cells (Finn 1996, 1998). That is, endometrial cells must proliferate and then differentiate, and once they differentiate, they have an expiration date. Ovulation and endometrial receptivity are fairly tightly timed, to the point that the vast majority of implantations occur within a three-day window (Wilcox, Baird, and Weinberg 1999). So it's not that menstruation expels dangerous menotoxins, but rather that menstruation happens because the endometrium needs to start over, and humans in particular have thick enough endometria that we can't just resorb all that blood and tissue.

It's time to dump the idea that menstruation is dirty. It's blood and tissue that you ended up not using to feed a fetus, and that's all.

REFERENCES

Clarke, John. 1994. "The Adaptive Significance of Menstruation: The Meaning of Menstruation in the Elimination of Abnormal Embryos." *Human Reproduction* 9 (7): 1204–07. PMID: 7848450.

Finn, Colin A. 1996. "Why Do Women Menstruate? Historical and Evolutionary Review." *European Journal of Obstetrics, Gynecology, and Reproductive Biology* 70 (1): 3–8. PMID: 9031909.

———.1998. "Menstruation: A Nonadaptive Consequence of Uterine Evolution." *Quarterly Review of Biology* 73 (2): 163–73. PMID: 9618925.

Freeman, William, Joseph M. Looney, and Rose R. Small. 1934. "Studies on the Phytotoxic Index II. Menstrual Toxin ('Menotoxin')." *Journal of Pharmacology and Experimental Therapeutics* 52 (2): 179–83.

Lemay, Helen Rodnite. 1992. *Women's Secrets: A Translation of Pseudo–Albertus Magnus'* De Secretis Mulierum *with Commentaries.* SUNY Series in Medieval Studies. Albany: State University of New York Press.

Martin, Emily. 1980. *The Woman in the Body: A Cultural Analysis of Reproduction.* Boston: Beacon Press.

Montagu, Ashley. 1940. "Physiology and the Origins of the Menstrual Prohibitions." *Quarterly Review of Biology* 15 (2): 211–20.

Perlstein, Meyer A., and Abe Matheson. 1936. "Allergy Due to Menotoxin of Pregnancy." *American Journal of Diseases of Childhood* 52 (2): 303–07.

Pickles, Vernon R. 1979. "Prostaglandins and Dysmenorrhea. Historical Survey." *Acta Obstetricia et Gynecologica Scandinavica* 58 (s87): 7–12.

Profet, Margie. 1993. "Menstruation as a Defense Against Pathogens Transported by Sperm." *Quarterly Review of Biology* 68 (3): 335–86. PMID: 8210311.

Reid, Helen Evans. 1974. Letter: "The Brass-Ring Sign." *Lancet* 1 (7864): 988. PMID: 4133673.

Wilcox, Allen J., Donna Day Baird, and Clarice R. Weinberg. 1999. "Time of Implantation of the Conceptus and Loss of Pregnancy." *New England Journal of Medicine* 340 (23): 1796–99. PMID: 10362823.

KATE CLANCY is an assistant professor of anthropology at the University of Illinois, Urbana-Champaign. Her research focuses on inflammatory processes of ovarian and endometrial function and how those influence life history trade-offs and human evolution. Her blog, *Context and Variation* (http://blogs.scientificamerican.com/context-and-variation), focuses on evolutionary medicine, evolutionary psychology, and ladybusiness. Kate is also a mother, a wife, a sister, a roller derby athlete, and a dog owner.

10

JOULE'S JEWEL

STEPHEN CURRY

For the longest time I thought he was French. It's the name—Joule; it *sounds* French, and in my physics class at school no one thought to explain otherwise. In fact, Joule was not even introduced as a name. The word was simply handed to us as the unit of energy, a replacement for Calorie, who, for all I knew, might also have been from France.

Many years later, long after I had completed my schooling and a degree in physics, I was surprised to discover that James Prescott Joule was from Salford, near Manchester. He remains a slightly elusive figure to me, but he is starting to take shape, more so since my latest visit to London's Science Museum.

It's a shame that histories and biographies are so often omitted from science classes, since the human story of how concepts developed can make them more accessible. This seems especially true of the more abstract ideas of science, which too many assume to have popped, ready-formed, from the mind of some genius. Joule was no genius; if anything, I get the impression that he was a bit of a plodder. But fortunately he was a tenacious, meticulous, and insightful plodder.

Joule's main interest was thermodynamics, which has a reputation for being difficult and boring, even among physics students who have opted to grapple with its abstruse state functions—temperature, enthalpy, entropy, and suchlike—and the seemingly endless differential equations that relate them to one another. But thermodynamics is more important than most of us (and most physics students) realize, because it connects our everyday experiences to the underlying atomicity of the world and helps us to make sense of "energy," a term much abused by the

(Photograph by Stephen Curry)

otherworldly but one that, thanks in part to Joule, has a well-defined meaning in science.

I was reminded of the Salford scientist by a chance encounter at the Science Museum with the apparatus that he used in his most famous experiment. I nearly missed it as I passed through the ground-floor gallery, on my way with my daughter to the IMAX cinema. But there, in a floor-level glass cabinet surrounded and dominated by the giant locomotives of the steam age, sat an unprepossessing piece of worked brass that had helped to extract the golden nugget of science that powers those engineered monsters. With that small canister and those rotating paddles, Joule showed the world that work and heat are equivalent. My daughter gazed patiently at the ceiling as, smiling, I flitted around the cabinet taking pictures.

So work and heat are equivalent; it doesn't sound like much. To most of us perhaps the notion is so obvious that any sense of its achievement has vanished. But eighteenth- and nineteenth-century science had struggled for a long time to grasp the imponderable quantities that were light and heat and electricity and magnetism.

Joule wasn't the first or the last to think deeply about them, but his careful experimentation was crucial to crystallizing our modern understanding. In part, he took his cue from Count Rumford, a delightfully colorful figure whose earlier cannon-boring experiments skewered the caloric theory of heat and pointed to the link between the force of friction and the generation of heat.

Like Rumford, Joule wrote an accessible account of his experiments for the *Philosophical Transactions of the Royal Society*, but the styles of the two men could not have been more different. Joule's report, "On the Mechanical Equivalent of Heat," which was published in 1850, has none of the egotistical exuberance of Rumford's rollicking tale. It is measured, plainspoken. But the ordinariness of the prose and the simplicity of the experiment are deceiving.[1] I think Joule's paper is a gem as lustrous as any worn by the preening count.

To begin with, Joule is fastidious and gracious in his acknowledgment of those whose work led science out of the blind alley of caloric theory to the point where he could conceive of an experiment to measure the equivalence of heat and work "with exactness." He mentions not only Rumford, but Davy, Dulong, Grove, Séguin, Faraday, and even the luckless Julius von Mayer (who, depressed by the lack of recognition of his work and in dispute with Joule over priority, attempted suicide in 1850).[2]

All of these men had been groping toward a realization that, in various ways, the application of forces to solid, liquid, and gaseous bodies caused them to heat up; and that heat was, in effect, the *energy* transferred to the body by the efforts of friction or compression or even electrification (which Joule linked to the *force* of chemical affinity in a battery). He even notes: "There were many facts, such as, for instance, the warmth of the sea after a few days of stormy weather, which had long been commonly attributed to fluid friction."

So the idea was there. But it was resisted. Joule notes at the end of his introduction that, despite the accumulating threads of evidence in favor of the relationship between work and heat, "the scientific world, preoccupied with the hypothesis that heat is a substance, and following the deductions drawn by PICTET [Marc-Auguste Pictet] from experiments not sufficiently delicate, have almost unanimously denied the possibility of generating heat in that way."

It was Joule's singular insight that *exact measurements* were the necessary foundation to convert the idea into a *scientific theory*, or better yet, into a law of Nature.

His apparatus was straightforward: a set of brass paddles immersed in water were linked to heavy weights by two pulleys on either side. As the weights dropped through a measured distance to the floor, the paddles rotated between fixed vanes and agitated the water. The weights were wound back up again and the process repeated.

The temperature of the water was measured at the beginning of the experiment and then again after twenty falls of the weights. The total increase in heat was calculated by multiplying the observed temperature rise—typically about 0.6°F—by the combined heat capacity of the water and the brass apparatus. The work done was determined by multiplying the combined weight (in pounds—lbs) of the two lead discs by the distance (in feet—ft) that they fell to the floor.

There is nothing terribly sophisticated about the experimental design, but the wonder of it is in the imaginative power and the heroic care that Joule applied to control and correct his measurements.

He engineered the apparatus to reduce the friction of any moving parts and fixed it on a wooden mount designed with as few points of contact as possible, so that any loss of heat through conduction would be minimized.

He shielded the apparatus from his own body heat with a large wooden screen, but even so, took care to do controls in which he went through the entire motions of performing the experiment but without moving the weights or the paddle, and measured the resulting temperature rise (which averaged at a minuscule 0.012975°F).

He measured the *room* temperature at the beginning and end of each experiment so that he could determine how any difference from the temperature of the apparatus was affecting the results (this required a correction of –0.000832°F).

He determined the *net* weight of the leaden discs (406,152 grains [grs], or 26.318 kg) by subtracting the frictional resistance of the pulley wheels of the apparatus, which he measured to be equivalent to a weight of 2,837 grs, an adjustment of 0.7 percent.

He measured the speed of the falling weights, which dropped at 2.42 inches per second, and subtracted the kinetic energy of their motion from his determination of the total work (weight times height fallen) to find the work that had been expended only on heating. Effectively, this entailed reducing the total distance dropped (1,260.248 inches) by 0.152 of an inch.

He repeated the experiment and each control run forty times.

And then he did it all again.

But the second time, he used mercury as the frictional fluid instead of water. This alteration required an additional step in which the vessel containing the mercury was immersed in water immediately upon completion of all the drops of the weights so that the heat accumulated could be measured by its effect on water. The mercury experiment, each time accompanied by a control run, was repeated twenty times.

And then he did it all again.

This last time, he mimicked Rumford by having the falling weights turn one disk of cast iron against another, while immersed in mercury. As previously, the whole apparatus was dunked in water at the end of each run to measure the heating effect of the friction. This was repeated, with controls, a sum total of ten times.

By three different experiments, using three different media—water, mercury, and iron—Joule found the heating effect of the work done by the falling weights to be the same value to within less than one-half of 1 percent. Consideration of the process and number of repetitions of each experiment persuaded Joule that the trials performed with water had provided the most accurate determination, and so he concluded: "That the quantity of heat capable of increasing the temperature of a pound of water . . . by 1° FAHR., requires for its evolution the expenditure of a mechanical force represented by the fall of 772 lbs through the space of one foot."[3]

The conclusion is prosaic, almost cumbersome. But through imagination, through meticulous care, through long, repetitious hours in the laboratory Joule pulled an idea out of the ether and gave it the substance to overcome the prejudices of men.

Thus did the first law of thermodynamics, enunciated by Mayer in the early 1840s but then resisted by many in the scientific community, even in the wake of the early reports of Joule's experiments, make its tortuous entry into the scientific mainstream. Mayer's rendition—let me

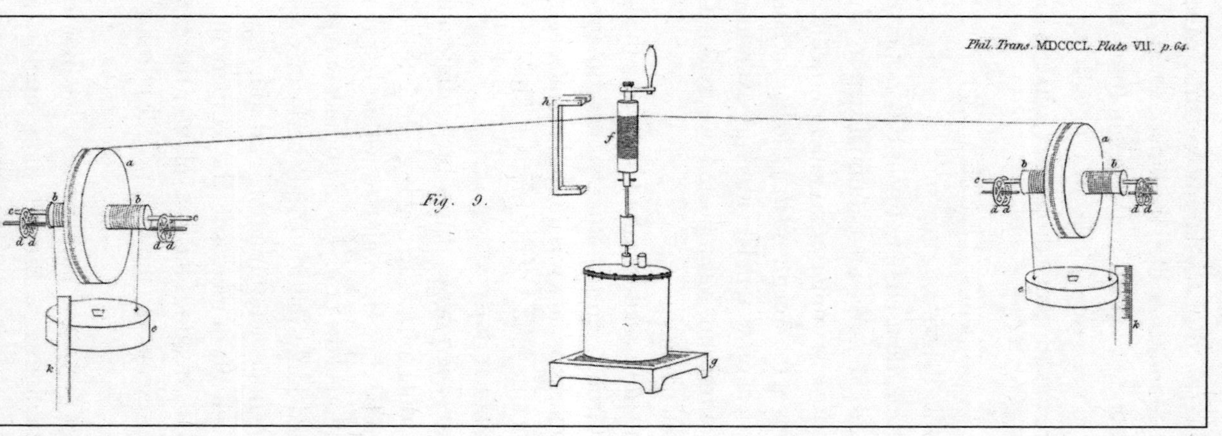

(Photograph by Stephen Curry)

give him the last word—is indistinguishable from the modern equivalent: "Energy can neither be created nor destroyed."

Our scientific understanding, in contrast, most definitely has to be created.

NOTES

1. The 1850 paper by Joule (*Philosophical Transactions of the Royal Society of London* 140: 61–82) is in fact a recapitulation of a series of experiments that he had started several years previously; his initial results were published in the *Manchester Courier,* which was an unusual destination even at that time.
2. Joule's words on Mayer appear to be especially carefully chosen, probably because his 1850 paper was written *after* the dispute with the German physician and physicist had erupted. He writes, "The first mention, so far as I am aware, of experiments in which the evolution of heat from fluid friction is asserted, was in 1842 by M. MAYER, who states that he has raised the temperature of water from 12°C to 13°C, by agitating it, without however indicating the quantity of force employed, or the precautions taken to secure a correct result." I haven't been able to get hold of Mayer's paper (published in *Comptes Rendus* 25: 421) but suspect there is some justification for Joule underscoring the point about the need for scientists to report the *details* of their experiments.
3. The result obtained by Joule is within 1 percent of the modern value.

See also John Gribbin, *Science: A History, 1543–2001* (London: Penguin Books, 2010), pp. 382–88.

STEPHEN CURRY is a professor of structural biology at Imperial College London. His main research interests are in RNA virus replication. On his *Reciprocal Space* blog at Occam's Typewriter (http://occamstypewriter.org/scurry/), he writes about his research, the scientific life past and present, and the role of the scientist in society. His writing has also appeared in *The Guardian*, the *Times Higher Education* magazine, and *The Biochemist*, and at lablit.com.

11

UNRAVELING THE FEAR O' THE JOLLY ROGER

KRYSTAL D'COSTA

The most recognized symbol of piracy is the Jolly Roger. The skull and crossbones flown from masts has long been a symbol of terror on the high seas, striking dread and despair into the hearts of those unfortunate enough to cross its path. This ominous design is an elemental part of the "pirate brand" and represents a magnificent exercise in collective hybrid branding.[1] That is to say, the icon is a versatile one—and remains recognizable despite any minor changes it may acquire. So powerful is this imagery that it persists today, but it was actually born out of an economic decision that united pirates under a version of the symbol to facilitate their intent to plunder.

The Jolly Roger is a fascinating example of the efficiency and power that good branding can deliver, but it also demonstrates the ways in which the power of a symbol is drawn in part from the acceptance and manipulation of the image by others.

Piracy has likely long been a feature of the open seas, following the earliest trade routes of the Aegean and Mediterranean. Cilicians were active in the Mediterranean and tolerated by the Roman Empire for the slaves they provided, and were reined in only when they gained such a presence as to become a threat to the Empire's grain supply in 67 B.C.E. The Senate approved "a comprehensive and systematic strategy and an astutely humane policy to the vanquished," in the words of the historian John L. Anderson, to eliminate the Cilicians within a matter of months.[2] Despite this historical legacy, the familiar skull and crossbones that many of us associate with piracy is a recent development, emerging in the late seventeenth century with the rise of the pirates of the Caribbean.

Following the exploration of the New World, the Caribbean quickly gained status as a center of trade, with sugar, gold, and human capital flowing between the Old and New Worlds. The Spanish dominated the landscape, but other colonial powers soon followed. Pirates, many of whom were drawn to the trade because it offered a chance to make a sustainable wage, found the waters of the Caribbean particularly attractive: the area was largely unsettled, so they would not be bothered by governing bodies; there were plenty of safe, natural harbors; and the brisk traffic afforded many opportunities to liberate spoils from the trade vessels of the Spanish.[3]

Tensions between Old World powers were not limited to their respective shores. Traces of these conflicts echoed in the Western colonies, and the English, Dutch, and French sanctioned piracy—commissioning pirates as privateers—as a means of protecting their claims and controlling the goods in the region. These men were national heroes—defenders of the nation on the high seas. Their numbers included Francis Drake and Henry Morgan, hailed as "gentlemen of the seas."

Pirates have a bloodthirsty and lawless reputation. They're known for making captives walk the plank, for copious alcohol consumption and lascivious tendencies. But these were skilled men, drawn from maritime trades that had paid them poorly:

> Merchant seamen got a hard, close look at death: disease and accidents were commonplace in their occupation, rations were often meager, and discipline was brutal. Each ship was "a little kingdom" whose captain held a near-absolute power which he often abused . . .
>
> Some pirates had served in the navy where conditions aboard ship were no less harsh. Food supplies often ran short, wages were low, mortality was high, discipline severe, and desertion consequently chronic.[4]

While privateers often had better food and pay and shorter shifts, the long arm of the law was sometimes unforgiving and held them to strict standards. Pirates who seemed to have no loyalties to man or country were able to set their own terms, albeit in the guise of crime. These seafaring groups were far from disorganized. They operated under strict codes of conduct that reflected a highly organized social order governing authority, distribution of plunder, and discipline. For example, spoils were systematically distributed: "Captain and quartermaster received between one and one-half and two shares; gunners, boatswains, mates, carpenters, and doctors, one and one-quarter or one and one-half; all others got one share each."[5]

The captain served at the mercy of the crew, and could be removed from his position for acts of cowardice, cruelty, or failure to act in the crew's best interest. A council governed the crew, representing the highest authority aboard the ship. In many ways this order was necessary to the survival of piracy. This group knew that they were operating on borrowed time and on the edge of the hangman's noose. Though they could be commissioned, if caught by an opposing party they faced death. They needed to hang together or they could literally find themselves hanging separately, and this bred a sense of fraternity that spread among pirates and manifested in cooperative tendencies at sea and in port. In this context, flags emerged as identifiers:

> In April 1719, when Howell Davis and crew sailed into the Sierra Leone River, the pirates captained by Thomas Cocklyn were wary until they saw on the approaching ship "her Black flag," then "immediately they were easy in their minds, and a little time after" the crews "saluted one another with their Cannon."[6]

Though conflict between pirate bands was not unheard of, the groups were largely cooperative, even across national boundaries. And they would defend one another. For example, when survivors of the wrecked *Whidah* were jailed in Boston in 1717, pirates soon captured a Boston ship captain, whom they told "if the Prisoners Suffered they would Kill every Body [the pirates] took belonging to New England."[7]

A version of the Jolly Roger was widely adopted by pirates for fraternal reasons that ultimately *did* lead to economic boons. Some 2,500 men sailed under a version of a black flag bearing the insignia of a white skeleton striking a bleeding heart with one hand and holding an hourglass. The flag was certainly meant to announce their presence, and the pirates, enterprising men that they were, quickly found that they could convey their intent to ships in their path with their banners. Black flags indicated that they were pirates but that they would consider providing quarter, while a red flag bearing the described insignia meant that no quarter would be given and the mates meant to fight to the end. However, the imagery chosen for the flag is as much a reflection of the pirates and their lifestyle as it was a reflection of their terrible natures:

> The flag was intended to terrify the pirates' prey, but its triad of interlocking symbols—death, violence, limited time—simultaneously pointed to meaningful parts of the seaman's experience, and eloquently bespoke the pirates' own consciousness of themselves as preyed upon in turn. Pirates seized the symbol of mortality from ship captains who used the skull "as a marginal sign in their logs to indicate the record of a death." Seamen who became pirates escaped from one closed system only to find themselves encased in another. But as pirates—and only as pirates—these men were able to fight back beneath the somber colors of "King Death" against those captains, merchants, and officials who waved banners of authority.[8]

By rallying under this sign, the pirates created a physical symbol that could be identified as "pirate." But perhaps this accepted branding suggested this group had grown too large and too powerful to be allowed to continue unchecked. Piracy was a business—an officially sanctioned business in many cases—but as the Cilicians were eradicated by the Empire once they became a sizable threat, so, too, would these men be persecuted, and by the very powers that once encouraged their numbers.

Historian Douglas R. Burgess Jr. discusses the ways in which the perception of seventeenth-century piracy was shaped by governing officials, who used the media for this purpose following the conviction of famed English pirate Henry Every. He was distinguished as a "noble pirate," a title also bestowed upon the likes of Drake and Morgan in recognition of their courage on behalf of English maritime interests.[9]

This reputation, having been firmly planted in the minds of the English population, proved difficult to undo. In fact, when Every was first brought to trial for the capture of the Indian treasure vessel *Ganj-i-Sawai*

("Gunsway") and the mistreatment of those on board, including women, he was acquitted by the jury, much to the embarrassment of the English government, which had taken the stance—somewhat necessarily to repair trade ties with India and restore the power of the East India Trading Company—that he must be punished.

On the national stage, this was a PR nightmare for England. The acquittal suggested that England was a "nation of pirates" to potential allies and trade partners such as India, and encouraged English colonies to sympathize with and support piracy in local waters because it suggested that the native England herself supported these individuals.

The government retried Every and his men under charges of mutiny. He had been first mate of the *Charles.* However, he seized the ship while at port—as he had not been paid—and renamed it the *Fancy*, and proceeded to attack the *Ganj-i-Sawai.* England effectively re-crafted the definition of piracy to bring him and his men to justice, and in doing so, sent the message that piracy itself would no longer be tolerated.

While this would not be the end of piracy, it may be a point at which multiple meanings associated with the Jolly Roger begin to take shape. The noble pirate image persisted: Every was treated as a folk hero in popular culture. For example, *The Life and Adventures of Captain John Avery*, published in 1709, painted Every as a "gallant swashbuckler who falls in love with an Indian princess on board the captured *Ganj-i-Sawai*," is elected king of a tribal island, and then decrees that his crew should also have "dusky" brides to share in his joy.[10] Subsequently, the playwright Charles Johnson would adapt the tale for the stage in *The Successful Pyrate*, which portrayed Every as an "empire builder" and a "tough but effective monarch."[11] Every was often subsequently depicted in military gear—the idea of the defender persisted.

The signing of the Treaty of Utrecht quieted much of the discord that had driven privateering initially, reducing the need for these seafaring brigands, as well as official tolerance for their actions.[12] In this context, signs of piracy became outlawed, to fit the idea that pirates are bloodthirsty, ruthless criminals. These ideas moved the symbols associated with piracy away from fraternal ties and self-identification, and moved the pirates themselves away from nationalist ties, rendering them targets of the state. And we are left with mixed symbolism—further diluted as it is appropriated for modern-day uses.

NOTES

1. Alice Rawsthorn, "Skull and Crossbones as a Branding Tool." *New York Times*, May 1, 2011. www.nytimes.com/2011/05/02/arts/02iht-design02.html.
2. J. L. Anderson, "Piracy and World History: An Economic Perspective on Maritime History." *Journal of World History* 6, no. 2 (1995): 175–199. www.jstor.org/stable/20078637.
3. Marcus Rediker, " 'Under the Banner of King Death': The Social World of Anglo-American Pirates, 1716 to 1726." *William and Mary Quarterly* 38, no. 2 (1981): 206. doi:10.2307/1918775.
4. Ibid., 207.
5. Ibid., 210.
6. Ibid., 219.
7. Ibid., 220.
8. Ibid., 223.
9. Douglas R. Burgess Jr., "Piracy in the Public Sphere: The Henry Every Trials and the Battle for Meaning in Seventeenth-Century Print Culture." *Journal of British Studies* 48, no. 4 (2009): 888. doi:10.1086/603599.
10. Ibid., 910.
11. Ibid., 910.
12. Ibid., 909.

KRYSTAL D'COSTA is an anthropologist who writes about the intersection of daily life, anthropology, and psychology at *Anthropology in Practice* (www.anthropologyinpractice.com), which is now part of the Scientific American Blog Network: http://blogs.scientificamerican.com/anthropology-in-practice. You can follow her on Twitter at @krystaldcosta.

12

FREE SCIENCE, ONE PAPER AT A TIME

DAVID DOBBS

On Father's Day four years ago, the biologist Jonathan Eisen decided he'd like to republish all his father's papers. His father, Howard Eisen, a biologist and a researcher at the National Institutes of Health, had published some forty papers by the time he died by suicide at age forty-five. That had been in February 1987, as Jonathan, a sophomore at college, was on the verge of discovering his own love of biology.

At the time, virtually all scientific papers were just on paper. Now, of course, everything happens online, and Jonathan, who in addition to researching and teaching also serves as an editor for the open-access, online-only journal *PLoS Biology*, knows this well. So three years ago, Jonathan decided to reclaim his father's papers from print limbo and make them freely available online. He wanted to make them part of the scientific record. He also wanted, he says, "to leave a more positive presence"—to ensure his father had a public legacy first and foremost as a scientist.

How hard could it be? Howard Eisen had been a federal employee, so his work rightly lay in some sense in the public domain. And Jonathan, as an heir, presumably owned copyright anyway, along with his brother, Michael (also a biologist, and one of the founders of the Public Library of Science, the innovative journal group that publishes *PLoS Biology*). Yet to the brothers' continuing chagrin, Jonathan has found securing and publishing his father's papers to be far harder than he expected.

For instance, even though Jonathan has access to the enormous University of California library system, which subscribes to a particularly

(Jonathan Eisen)

high number of journals, he often can't even find some of his father's papers. And when he finds a paper in a journal the university doesn't subscribe to, he is asked to pay as much as $50 to read it—even though his father did the work with public funds. He's not alone; one recent study found that even most university researchers have access to only about half the papers they need to cite for a given bit of research. At one point Jonathan even asked on Twitter if anyone could send him a copy of one of his father's papers; someone tweeted back that they'd tried but hit a paywall, and jokingly asked for his credit card number. "I ain't payin'," he replied.

Meanwhile, Jonathan found and downloaded the PDFs for about half his father's papers, but he remained uncertain whether he could safely post them on his website. While some publishers allow such "collegial sharing," others leave their policies unclear, and he worried about getting sued. His brother urged Jonathan to post them.

"Come on," Michael wrote in a comment at Jonathan's blog. "I DARE them to sue us."

Jonathan posted the whole list on his blog and uploaded what PDFs he could obtain, and so far he has not been sued or asked to take them down. Yet he remained wary and unsatisfied. He knew that few researchers would find his father's work if the papers reside only on his web page. So for the time being, his father's work stayed stuck in an old structure—a calcified matrix. Though Jonathan banged away at the surrounding rock, he knew he hadn't pried the work loose. This frustrated him on two fronts: It kept him from freeing his father's work. And it confirmed to him that much of the content of science, which should be a fluid medium, is still trapped in old structures.

"I started this partly to test how hard it would be to try to make science more available in the current system," he says. "I'm finding that even with my father's papers, or even with my own, it's not very easy."

Jonathan Eisen's quest has solidified his conviction that science needs to radically rework the way it collects and shares its data, methods, and findings. He has plenty of company. A growing number of prominent scientists want to replace the aging journal system with something faster, cheaper, and richer. The current system, they note, grew out of meeting notes and journals published by societies in Europe over three centuries ago. Back then, quarterly or monthly volumes could accommodate the flow of ideas and data from most disciplines, and the printed journal, though it required a top-heavy, expensive printing and publishing infrastructure, was the most efficient way to share those ideas.

"But now," says Jonathan Eisen, "there's this thing called the Internet. It changes not just how things can be done but how they *should* be done."

As the Stanford biochemist and Public Library of Science (PLoS) cofounder Patrick Brown put it a few years ago, "What seemed an impossible ideal in 1836, when Antonio Panizzi, librarian of the British Museum, wrote, 'I want a poor student to have the same means of indulging his learned curiosity, . . . of consulting the same authorities, . . . as the richest man in the kingdoms,' is today within reach. With the Internet, we have the means to make humanity's treasury of knowledge freely available to scientists, teachers, students and the public around the world."

"The existing system worked well for quite a while," says Jonathan

Eisen. "But it was not designed by theory. It was designed by constraints." In a world that provides communications conduits far larger and faster, those constraints have now made science's traditional pipeline a bottleneck.

To get a sense of how the current system curbs science, consider a rare case in which researchers attacked a big medical problem with an open-science model. In 2004, in the United States, a network of government and private researchers, including large drug companies, used open-science principles to accelerate research into Alzheimer's. The project, as Gina Kolata aptly described it in a 2010 *New York Times* article, "was an agreement . . . not just to raise money, not just to do research on a vast scale, but also to share all the data, making every single finding . . . immediately . . . available to anyone with a computer anywhere in the world." Before that, researchers had worked separately, siloing off much of their work. Now methods and data formats were standardized. The data would immediately enter the public domain, where anyone could build on it.

An extraordinary project ensued. The U.S. National Institute on Aging contributed over $40 million, and twenty companies and two nonprofit groups kicked in another $27 million to fund the first six years. The program produced an explosion of papers on early diagnosis and helped generate more than 100 studies to test drugs or other treatments. It greatly sped and opened the flow of findings and data. According to *The New York Times*, the project's entire massive database had been downloaded more than 3,200 times as of August 2010, and the data sets containing images of brain scans have been downloaded more than a million times. Everyone was so pleased with the results that they renewed the accord the following year. And all because, as a researcher told Kolata, "we parked our egos and intellectual-property noses at the door."

The language used here—everything entering the public domain, the dismantling of silos, the parking of egos and intellectual property padlocks—might have been lifted from an open-science manifesto. And even Big Science appreciated the outcome. To open-science advocates, this raises a good and somewhat obvious question: why don't we do science like this all the time?

Part of the answer, strangely, is the very thing at the center of science: the paper. Once science's main conduit, the paper has become its choke point.

It's not just that the paper is slow, though that is a huge problem. A researcher who submits a paper to a traditional journal right now, for instance, won't see the published piece for about a year. She must wait while the paper gets passed around among editors, then goes through rounds of peer review by experts in her field, who might and often do object not just to her methods or data but to her findings and interpretations. Finally, she must wait while it moves through an editing, layout, and publishing pipeline that itself might run anywhere from two to twelve weeks.

Yet the paper is not simply slow; it's heavy. Even as increasingly data-rich science has outgrown the paper's ability to deliver and describe all that science has to offer—its deep databases, its often elaborate methods—we've loaded it up needlessly with reputational weight and vital functions other than carrying data.

The paper is meant to be a conduit for the real content and currency of the science: the ideas, methods, data, and findings of the people who do science. But the tremendous publishing and commercial infrastructure built around the academic paper over the last half century has concentrated so many functions and so much value in the journal that the paper itself, rather than the information in it, has become science's main currency. It is the paper you must buy; the paper you must publish; the paper you must cite; the paper on which not just citations but tenure, reputation, status, and even school rankings are built.

To get an idea of the paper's excess weight, go to Cambridge, England, and find Mark Patterson. Patterson is a scientific-publishing old hand gone rogue. He formerly worked at two of the biggest scientific publishing companies, Elsevier and Nature Publishing Group (NPG), each of which puts out scores of journals. A few years ago he moved to the staff at PLoS.* Patterson is now director of publishing there, and since he

* Disclosure: I sometimes write for Nature Publishing Group and have friends both there and at PLoS.

joined, PLoS has leveraged open-science principles to become one of the world's biggest publishers of peer-reviewed science and the biggest single publisher of biomedical literature. Readers like it because they get free access to good science. Researchers like it because their work reaches more readers and colleagues. PLoS's success is heartening open-science advocates greatly—and unsettling the traditional publishers.

To describe PLoS's innovations, Patterson likes to talk about how PLoS's most innovative journal, *PLoS One*, deals with four essential functions of science that are currently wrapped up in the scientific paper: registration, certification, dissemination, and preservation. The current publishing regime, he argues, locks up these functions too closely in the current, conventional version of the scientific paper—even though some of these functions can be met more efficiently by other means.

So what are these functions?

Registration is essentially a scientific claim of discovery—a marker crediting a particular researcher with an idea or finding. The current system registers these contributions via a paper's submission date. *Certification* is essentially quality control: ensuring a paper is solid science. It is traditionally done via peer review. *Dissemination* means getting the stuff out there—publication and distribution, in printed journals or online. And *preservation*, or archiving, involves the maintenance of the papers and citations to create a bread-crumb trail other researchers can later follow back to an idea or finding.

"The current journal system does all four of those things," says Patterson. "But it doesn't necessarily do them all well. The trick is finding a system that gets each of these done most efficiently, sometimes by other means, instead of having them all held by the publisher." He and others contend that science would gain both speed and rigor by "unbundling" some of these functions from the paper and doing them in new ways.

PLoS loosens things up mainly in distribution and quality control. All of its journals are open-access—that is, free to read. Instead of making every would-be reader either buy a journal subscription or pay a per-article price of $15 to $50, PLoS collects a fee from the researcher to publish—usually about $1,400 or so—and then publishes the paper online and makes it free. The author fee is substantial, but it's actually a small addition to the other costs of doing science, and performs the essential function of getting it out there. It's Panizzi's dream realized: every poor schoolchild—or at least every schoolchild with web access—can read PLoS journals. Researchers like this, and it works. A recent study showed that on average, papers and data published open-access receive more citations than do those behind paywalls.

PLoS's rapid growth has shaken things up. Some journal groups, such as Elsevier, have responded by allowing authors to pay to have a paper open-access on publication. Yet commercial publishers that do this tend to retain certain rights that PLoS does not, and they're less likely to release underlying data, metadata about the publications, or other data and rights. And the practice creates a weird and uncertain market: you can go to, say, *Neuron*, and find, in the same issue, one paper you can download for free and another that costs $30. The difference? The authors of the latter paper didn't pay the open-access fee.

Meanwhile, PLoS's biggest, most cross-disciplinary journal, *PLoS One*, streamlines quality control in a way that's more complex and raises more ire. The traditional route, peer review, generally involves having two or three experts evaluate the entire paper—data, methods, findings, conclusions, significance. The publisher relays these peer critiques to the author, usually with requests for either changes or clarifications. If the author answers those to the publisher's satisfaction, the paper gets approved.

PLoS One uses a similar process but—crucially—asks its reviewers to judge only on technical merits, and not on any assessment of the paper's novelty, significance, or impact. "The idea," says Patterson, "is to let the importance be determined later by how much the paper's ideas and findings and conclusions are taken up by the community. We're letting the scientific community at large determine a paper's value and importance, rather than just a couple of reviewers."

This makes many people at Patterson's old workplaces uneasy. Gerry Altmann, editor of *Cognition*, an Elsevier journal, and an open-minded man, doubts this sort of post-publication filter can serve the purpose. "Peer review *should* be about ensuring that there's a robust fit between findings and conclusions, and that a paper sits well within the context of a discipline," Altmann told me. "These are insidious changes."

Can the hive mind do quality control? Patterson answers by noting that any paper's true value—its lasting contribution—is generally decided by the scientific community even under the current system. Yet he acknowledged that at present few scientists actually go online and make comments or otherwise review papers published there. We're a long way from the vision of an active scientific community replacing peer review with a crowdsourced rigor and fact-checking. The hive mind apparently has better things to do. Altmann thinks it's starry-eyed to think that will change.

Others say researchers would engage these tasks if it was worth their while. They argue that you can make it worthwhile by giving researchers credit for a wider range of contributions to science, starting with post-publication peer review and evaluations.

This is the idea behind ORCID, a program that would give each researcher a unique, immutable digital identification, somewhat like a permanent URL. That ID would serve like a deposit account: the researcher would accumulate reputational credit not just for papers published, but also for other contributions the scientific community deems valuable. Reviews of others' work could thus generate deposits, as could public outreach, talks, putting data online, even blogging—anything that helps science but currently goes unrewarded. This would allow hiring, tenure, grant, and awards committees to weigh a broader set of contributions to science. ORCID holds particular promise because it has already lined up buy-in from publishing giants Nature and Thomson Reuters (though it's unclear what contributions various stakeholders will agree to credit).

What would such a system look like? One idea is being developed by a team led by Luca de Alfaro, of the University of California, Santa Cruz. Working with Google, the team hopes to develop broader-based reputational metrics that are built, writes de Alfaro in a recent essay in *The Scientist*, "on two pillars": tenure, grant, and similar rewards for authors of papers and their reviewers alike; and—crucially—a content-driven way of gauging the merit of both papers and reviews. Authors would get credit for work of high value, as measured by citations, reuse of data, and discussion generated. Reviewers, meanwhile, would get credit based not just on output but on how well their reviews predicted a work's future value.

"Thus two skills would be required of a successful reviewer," writes de Alfaro: "the ability to produce reviews later deemed by the community to be accurate, and the ability to do so early, anticipating the consensus. This is the main factor that would drive well-respected people to act as talent scouts, and to review freshly published papers, rather than piling up on works by famous authors." De Alfaro says much of the technology to weigh such variables already exists in algorithms used at Google and (for evaluations of reviewers) Amazon.

Such a system could readily be incorporated into a program like ORCID. It could also give researchers incentives and credits—points, essentially—for public outreach or for openly sharing underlying data and details about method, both after and even before publication, so that other researchers can more easily test or use the data and methods. In short, a more flexible credit system could generate more activity in almost any area of science simply by weighting it more heavily.

These many pressures and alternatives seem to be loosening the publishers' grip. In June 2010, librarians at the University of California system balked when the Nature Publishing Group sent a contract renewal containing a 400 percent price hike on the scores of NPG journals the huge library system subscribes to. The increase would have jumped the cost to over $17,000 per journal. The librarians objected that it was ludicrous for universities to fund research and then pay to read it. They threatened to boycott NPG not just as subscribers but as contributors to the journals. NPG softened and worked out a deal.

Meanwhile, universities and researchers are rebelling in other ways. Some are starting open-access journals or opening up some they already publish. And PLoS continues to create new models, including fast-track journals for time-sensitive disciplines, such as those that cover the flu and other infectious diseases, to cut the traditional one-year publication cycle down to a day. Another outfit, LiquidPub, is launching what it calls "liquid journals," in which "social computing and liquid knowledge will shape and navigate information waters." Phillip Lord and Robert Stevens, of Newcastle University and the University of Manchester in the U.K., have created Knowledge Blog, a publishing framework based on blog technology. Even Shakespeare scholars are entering the open-science world: in 2010 the *Shakespeare Quarterly* ran an experiment in which it not only put its journal online but opened the job of peer review to the public, so that anyone who cared to register could comment, say, on the racial implications of playing Titus Andronicus as an "American Gangsta."

And then there are those such as Phillip Lord, mentioned above, who just publishes on Wordpress. A blog may seem a sketchy way to publish science. Yet in a way it makes sense. Science, however rigorous, implicitly recognizes that every explanation is provisional; there's no finished version. So what could be more fitting than to revamp science through a platform explicitly built to be revised, commented on, and updated?

Yet if a more open scientific publishing landscape may seem inevitable, it's hardly clear how to get there. Talk of inevitability hasn't much helped Jonathan Eisen get his father's papers out in the open. He has struggled to find the right leverage point, or perhaps the right tool, to lift them onto a platform any more prominent than his own web page.

Lately, however, Jonathan has been using a tool that adds some leverage—and just might chip away at the calcified matrix in which his father's and others' scientific work has been stuck.

It is the simplest of academic tools: a desktop reference manager called Mendeley. Yet it comes with an extra dimension: a website at which you can share papers you like, creating a metadata-rich index that can lead other users to your user profile and papers, and vice versa. You load your bibliography up—all those papers on cognitive neuroscience, say, or dark energy, or, if you're Jonathan Eisen, evolutionary biology and extremophile bacteria—and Mendeley's algorithms link you up with papers you might have overlooked and the researchers who wrote, read, or collected them.

Maybe, Jonathan wondered aloud on Twitter, he could create a posthumous profile for his dad and post his papers there. Mendeley promptly told him he could. He did. Howard Eisen now has his own Mendeley page, with all forty-one of his papers listed and twenty-four of them uploaded as PDFs. Now you, as well as anyone in the research community who takes a minute to sign up at Mendeley, can find and read them, and add them to your library. Since Mendeley now has well over a million members and is growing at an accelerating rate, this puts Howard Eisen and his work if not in science's mainstream, then in a sizable and fast-growing tributary.

Jonathan also likes Mendeley because it seems to advance the larger open-science agenda. It's a sort of friendly Trojan horse. You download a reference manager—a good one, and free—and suddenly have a tool that can help open science.

"Smart," says Leslie Carr, a director of the Web Science Doctoral Training Center at Britain's University of Southampton. "Most people who've tried to create software to drive open science have started off on the web and tried to encourage people to share. Mendeley starts where the researchers already are, with a tool researchers need, which is a desktop reference manager to manage their bibliography and organize their thoughts. Then the very act of looking for more papers leads them into an open-science model based on sharing."

Mendeley exists because its CEO and co-founder, Victor Henning, needed a tool to understand better the cross-disciplinary mountain of literature he'd compiled for his thesis at Bauhaus–Universität Weimar. "I had this huge trove of papers from different disciplines," Henning told me, "and I wanted to see where the connections and overlaps and gaps were." But when he looked for software to do this, he found nothing he liked.

"This was 2004, 2005. Last.fm [an online music-sharing and recommendation service] was happening. By then I was collaborating a lot, and we were talking about doing a couple different people's data. Then we realized you could do a social version of this. Why not map a *bunch* of people's data? That would give an even better picture of the ideas in play." By this time, being a business student, he was thinking: startup. He also realized he didn't like most of the reference managers on the market. So he thought: let's roll that in, too.

Thus emerged Mendeley. The name rose from its dual mission: Mendeleev was the Russian chemist who created the periodic table, which organized the known elements into a structure that suggests the properties of other elements still to be found. Mendel was the nineteenth-century monk and botanist who saw that crossing two packets of information could yield a third packet that derived but differed from the first two. Two nice models of how science works.

The focus on the paper came of pure necessity. The American bank robber Willie Sutton said he robbed banks because "that's where the money is." At least for now, the paper is where the data are. But the people at Mendeley know quite well that (a) the paper will get unbundled and in many functions displaced, and (b) they're now grasping a bundle with a bunch more stuff in it. But they're most interested in the threads that run from paper to paper. They mean to charge not for the bundled information but for helping people find the connections between the bundles.

The company offers a free version that accommodates smallish libraries. If your library runs bigger than 500 MB, you can pay $10 a month to run the company's algorithms and store a copy of your data and papers in the cloud. If you're a company or a department or simply someone who wants to run some highly sophisticated or customized analyses on aggregated scientific data—and on the all-important hive mind indications of what's newly hot—you can pay more, providing Mendeley another income stream. Mendeley also talks of striking a deal, maybe, with publishers to make papers available on a rough iTunes model: a buck a paper, perhaps, with algorithms running in the background to help you find papers you'd like but don't know about.

Many feel this model holds a lot of potential not only to make papers available more freely (or cheaply), but to help unbundle and redistribute the functions now unnecessarily bundled with the paper. But can Mendeley do this? It seems to possess the vision and flexibility of mind. While Mendeley is necessarily focused on the PDF right now, for instance, there's no reason it can't adjust its databases and algorithms to index, share, and analyze the importance of contributions other than traditionally published papers; they can do new metrics. And the company's advanced programming interface, or API, recently published, should allow outside developers to create modules and add-ons to track new metrics, including author identifiers such as ORCID.

The chassis, then, can accommodate changes under the hood. The trickier part may be getting the steering right—that is, creating a user interface that offers a powerful and full-featured but easy and intuitive way to use both the traditional reference manager and the broader social, sharing, and analytical tools.

They're still working on that. "Most of the people I talk to who've used this," says Leslie Carr, "think that the desktop and the web sharing aren't as well integrated as they could be, from a software perspective, and that the analytic tools aren't as accessible or transparent as they should be." I find the same thing myself: many of the metrics and connections between papers aren't accessible on the desktop, presumably because they require the server's data and processing power, and finding them on the web interface feels vaguely opaque. Even when you find some relationships, you worry you're missing something.

Yet the company seems both open and responsive. When users pressed for more hive mind information and more fluid sharing, the company substantially upgraded the website's social-sharing module. It was quick to produce iPhone and iPad versions of its software. In general it appears to be fairly nimble and eager to meet user needs.

On the other hand, a lot could stop Mendeley. It could run out of money: with over forty employees, its burn rate is high. But then again, its funding angels seem both confident and deep-pocketed. It could get sued. It could fail to add features fast enough to satisfy demanding users. It could not *quite* create the magic that software needs to be transformative. In short, it'll need what any game changer needs: a good concept, some serious programming and promotional chops, and luck.

Mendeley chief scientist Jason Hoyt thinks the real killer app in open science will not be software but . . . the researcher. In August 2010 he made the call in a blog post titled "Dear researcher, which side of history will you be on?"

For the past three centuries, he noted, technology has prevented us from fulfilling Panizzi's dream of fast, free science. But the technology is there now, and so are the business models, as PLoS has shown. So what is the revolution waiting for?

It is waiting, wrote Hoyt, "for us, the researchers":

> We could choose to publish in only Open Access. We could choose to reward tenure for Open Data. We could choose to

> only reward publications or data that are proven to be reused and make either a marked economic or research impact. Instead, we choose to follow a model that promotes prestige as the primary objective . . . History, I suspect, will look upon our society and practice with regards to scientific knowledge-share as we similarly do now with the Dark Ages. **Each time we hold back data or publish research that isn't immediately open to all, we have chosen to be on the wrong side of history.**

He has a point. It's interesting, for instance, to imagine what would happen if researchers and university librarians got together and created a global version of the sort of revolt that the University of California librarians threatened. "You get all the librarians together on this," says Cameron Neylon, a director at the U.K.'s Science and Technology Facilities Council and an academic editor at PLoS, "and this is pretty much over." And *Librarians at the Ramparts* sure makes a nice image.

Jonathan Eisen, too, thinks that opening science will require the researchers to step up. But he suspects they won't step up in number until reward systems offer some incentive more tangible than being on history's good side. Only then will the upslope ease. In the meantime, Jonathan continues to push his father's papers up that hill, and he waits to see how well Mendeley, among his other efforts, can help pull them up into the open. Jonathan tends to push hard in strong spurts around Father's Day, make some progress, then set the load down a while before resuming.

"It's one of those things that's just going to take some time," he says. "I didn't think it would be quite so hard. But we'll get there."

Note: A week or so after this essay was published on the author's *Neuron Culture* blog at Wired.com, Jonathan Eisen, buoyed by the attention it generated, made a final push and finished posting all of his father's publications at Mendeley.

DAVID DOBBS writes for publications including *National Geographic, The Atlantic, The New York Times Magazine, Wired, Nature, The Guardian,* and *Scientific American.*

He is currently writing a book with the working title *The Orchid and the Dandelion*, which explores the notion that the genes and traits underlying some of our most vexing mood and behavior problems may also generate some of our greatest strengths, accomplishments, and happiness. The author of three previous books about the environment, science, and culture, he blogs on these and other subjects at *Neuron Culture* (www.wired.com/wiredscience/neuronculture/). Find him at http://daviddobbs.net/.

13

TINEA SPEAKS UP: A FAIRY TALE

CINDY M. DORAN

Once upon a time there was a national meeting of fungi, overseen by fairies and witches to keep it running smoothly. The fungi were gathered from all over the world to determine who should rule their kingdom. The main criterion for ruler, oddly enough, was how well a fungus could convince the other fungi of their capacity to affect the human species.

Mushroom spoke with authority; he believed he should be ruler. All the other attendees squirmed in their circle, especially the slime molds.

They didn't like the fact that they had all recently been divided up and kicked out of the Fungi kingdom. They were here out of protest. Needless to say, the gathering was an uncomfortable fairy ring of sorts. Mushroom had always talked way too much about his pharmaceutical powers. Most fungi couldn't listen to him anymore.

Tremella of Lincolnshire, a jelly fungus, shook off a few fruit bodies in disgust. "Really now, Morel, I should be ruler, as many humans believe I am from the stars."

Less cautious rumblings arose from the crowd. A few fungi of other species nodded either their caps, or conidia, or bud scars, or sporangia, whatever was up where a head might be (they were diverse and in various states of development). It was true; none of them were called "starshot" like *Tremella* was.

A little gray-green lichen from County Mayo murmured that graveyard mold held the most power. "Graveyard mold has been used at human funerals to keep the dead from reappearing. I think it should be ruler. Who else here can claim such power over humans?"

(Mycetozoa—slime mold—from *Kunstformen der Natur* [1904], plate 93: print by Ernst Haeckel [1834–1919].)

Now this once-calm meeting was more like a crowd at a Germany vs. England World Cup soccer match. No one saw the microscopic dermatophyte self-proclaimed as *Tinea* (she is slightly egotistical) climb onto the oak stump stage. She spoke into the microphone: "Ahem . . ."

No effect.

She was patient and slowly caught everyone's attention. She was used to being ignored at first, but she knew that if she persisted, others would eventually listen. After five minutes of "Ahems" the crowd silenced and squinted.

(Photograph by Curtis M. Andrews)

"Who's up there?"

"Ah, it must be another one of them tiny fungi. Perhaps we should listen," choked *Aspergillus*.

The meeting referees flew over the gathering and asked for all the fungi's attention.

Tinea capitis (a disease name she personally took as her own because it was her favorite) stood there, with hyphae out, proud and calm. She read from notes that no one else but she could see. "I stand before you a humble *Trichophyton*. I don't claim to have magical powers, but medical powers I have. I cause many skin infections, but ringworm of the head, *Tinea capitis*, that is"—she blushed—"has had the greatest effect on the social ordering of humans." Her view was obviously biased and unfounded, but other agents of infectious diseases, such as malaria parasites, retroviruses, bacteria, and the like, were not at this meeting.

Upon the words "social ordering," a communal inhalation of forest air brought all movement to an end. The thinned air hung over the birds that had stopped their nesting and were peeking over the rims of their homes.

Everyone was paying attention.

Evoking the power of three, *Tinea* set forward her claim. "I have the power to affect human social ordering in three ways." A royal blue slide with tiny gold print showed her points. It wasn't readable at the far side of the circle. She read them off. "Number one: I can spread fear. This makes humans listen to and agree to practically anything. Number two: I can make beautiful people invisible."

Gasping and murmuring interrupted her.

Tinea patiently proceeded. "Number three: Some of the most hideous scalps are upon the most clever boys—the ones mistaken for stupid. Alone and discredited their whole lives, they become heroes." She continued. "Please let me explain my position by providing some examples of each point. All humans read the newspaper, and I love the power of the press, so I'll take stories reported in *The Daily Tale* to illustrate my points."

She saw that the crowd was listening and went on, becoming more animated. "In 1956, Italo Calvino highlighted old tales of the danger of the mangy one. In 'The Ship with Three Decks,' a young lad (the true godson of the King) set upon a journey to meet the King. Before he set off, he was warned of three individuals, one with the mange! The third man he met on his journey was the mangy one, but he was disguised by a wig. The clueless lad traveled with the wigged one, who stole his godson-identity. The lad was forced to be his servant. When they reached the King, the King believed the wigged one to be his godson.

"Now, the wigged-mangy one, being truly evil, wanted to keep his identity a secret. So he sent the lad on a dangerous journey to rescue the King's daughter—hoping that the lad would be killed. At the end, the mangy one was found out, and the lad rescued the daughter. What is important to glean from this story is that evil ones with mange give ringworm, *me*, a bad name. This bad name leads to fear of all mange, whether it's me or not. That gives me power.

"Joseph Jacobs had a wonderful report in *The Daily Tale* called 'The Three Heads of the Well' where a cruel humpbacked stepdaughter of the King of Colchester was punished for being cruel to an old man and three heads she found in a well. She was given the mange on the face. Now, I know that this was truly leprosy (an infection that affects the colder parts of the body, not the cap of the head), but the term 'mange' associates me by default—increasing the fear associated with all 'mange.' This fear is so pervasive that the term 'mangy' is clearly an insult in human circles.

"Well, one reporter, Perrault, was too tame in his telling of 'Puss in Boots' to use the term, but Giambattista," she stretched the vowels as long as she could, "Giambattista Basile, as told by Peters, got it right when the cat assured the miller's youngest son that he wasn't 'worth less than a mangy donkey'—the inheritance of the second oldest son."

She showed pictures of heads with patches of hair missing, skin all scaly and lumpy. "I cause scalp disease, and am commonly confused with other diseases that may affect the scalp. Allergy, lupus and other autoimmune diseases, psoriasis, seborrhea, even leprosy." Eyes rolled upward.

"Other infections are also confused with me. I am truly mostly ringworm (a treatable disorder), but put me in a crowd of people and I affect everyone, whether they have me on their heads or not. Scald head. Scalp disease. The mange. Whatever the cause, the fear leads to this ignorant reaction among humans—social division based on appearance."

This was news to the fungi, as they were incredibly diverse, and used to the idea of being different from one another. *Tinea* let this sink in for a moment, and continued.

"For a long time now, people with scald head have been ignored. Beautiful people and royal people have noticed this, but have used it to their own advantage. I know of two reports from two different countries of young golden-haired boys from royalty who hid their identity by covering their heads, one with a cap he claimed was to hide a sore (as Grimm reported in Germany in 'Iron Hans'), the other with an ox bladder to make his head look like it was covered with the mange, as Calvino reports from Italy in 'The Mangy One.' One boy was taken from his family by the devil, and one by a forest wild man. After masquerading as gardeners' helpers with the mange, they somehow won the hearts of princesses who guessed their true identities. In the end, the golden-haired ones married the princesses and reunited with their royal families. They hid for so long, however, because of me!"

She puffed out more spores and told of how some poor people have less sanitation and were unable to get rid of their mange. Because of this mange, they became further discredited—almost invisible within society. She knew the humiliation. These people had an inner power, however, that made them better than others—because of her.

She spoke of the *Arabian Nights*. "On the seven hundred and seventy-eighth and seven hundred and seventy-ninth nights, Shahrazad told the story of the scald-headed boy who spied on the Fellah's wife, in 'The Fellah and His Wicked Wife.' The wife was sloppy around the boy—taking him for dumb. You see, the wife was having an affair, wanted her husband gone, and tried to poison him. The boy saw what was going on and dutifully informed the Fellah, who took corrective action. She should never have dismissed the boy. He was a clever lad.

"Calvino also reported on a clever lad who was the son of a cobbler in 'Quack, Quack, Stick to My Back.' He, too, had scalp disease. He wished to win the hand of the depressed princess by making her laugh. If he tried and failed, he would be put to death. His family laughed at him: 'Ha, ha, ha . . . a king with scalp disease.' When he left home for his mission, he was quoted in *The Daily Tale* as saying, 'Well, father, I'm leaving. Here everybody looks down on me because of my scalp disease. Give me three loaves of bread, three gold florins, and a bottle of wine,' and he went. Along his way, he fed three poor women out of the goodness in his heart. The third woman, after the last of the items was gone, turned into a blond maiden with a star in her hair, and she gave him a fine goose with instructions: 'If touched, the goose will say "Quack, Quack" and you should promptly say "Stick to my back."'

"At an inn he was offered free room and board because the innkeeper's daughters eyed the goose. In the middle of the night the goose said, 'Quack Quack,' to which he replied, 'Stick to my back.' The daughter, who was trying to steal the goose, became stuck to it. She yelled for her sister, who came in her nightgown and became stuck in the same manner.

"Evidently nonplussed, the boy continued his journey along with the goose and stuck daughters. In a timely fashion, a priest and coppersmith with pots and pans became stuck, too, just a short way from the castle of the depressed princess. She was on her balcony, saw the sight, and laughed—as did everyone else in the castle. The boy asked for the daughter's hand in marriage, but the King didn't want to give permission because of the boy's scalp disease. The lad didn't take 'No' for an answer, so the King ordered him bathed. Of course, he became handsome and married the princess.

"In 'The Golden Goose,' Grimm reported the same story, but called the boy a simpleton. The lad was kind and clever (not a simpleton). He won the princess, in a slightly different manner. The omission of him being a scald-head was a major journalistic error!"

She concluded her speech with a final puff of spores, waving them over the crowd. Shiitake and others placed leaves on their head-like structures to protect themselves. "Who here can hold such power?"

The vote was unanimous. *Tinea capitis* truly had powers of epic human consequence.

To bed now. To bed. Go rest your weary head.

NOTES

In reality, ringworm is caused by a *Trichophyton* that causes skin and hair infections. Other dermatophytes can cause these, plus nail infections. A dermatophytic infection of the scalp is called *Tinea capitis*, a treatable affliction no longer associated with the stigma, in the developed world, that it had at the time of the Arabian Nights, Basile, Grimm, Jacobs, and even Calvino. *Tinea*'s opinions are hers alone, not those of Cindy M. Doran.

A very heartfelt thank you to Jack Zipes for leading me to "Iron Hans" (Grimm) and the main reference for fairy tales: Uther's three-volume *The Types of International Folktales*.

REFERENCES

Arabian Nights, in 16 Volumes—Vol. XV, Supplemental Nights to the Book of the Thousand Nights and a Night. Translated and annotated by Sir Richard F. Burton. New York: Cosimo Books, 2008. First published 1888 ["The Fellah and His Wicked Wife"]. www.wollamshram.ca/1001/Sn_5/15tale14.htm.

Calvino, Italo. *Italian Folktales*. Translated by George Martin. New York: Harcourt Brace Jovanovich, 1956 ["The Mangy One"; "Quack, Quack! Stick to My Back!"; "The Ship with Three Decks"].

Duggan, Frank M. "Fungi, Folkways and Fairy Tales: Mushrooms and Mildews in Stories, Remedies, and Rituals, from Oberon to the Internet." *North American Fungi* 3, no. 7 (2008): 23–72. doi:10.2509/naf2008.003.0074.

Jacobs, Joseph. *English Fairy Tales*. Everyman's Library Children's Classics. New York: Alfred A. Knopf, 1993. First published 1890 ["The Three Heads of the Well"]. www.authorama.com/english-fairy-tales-46.html.

Owens, Lily, ed. *The Complete Brothers Grimm Fairy Tales*. New York: Avenel Books, 1981 ["Iron Hans"; "The Golden Goose"].

Peters, Amy. *The Everything Fairy Tales Book*. Topeka, Kan.: Sagebrush Education Resources, 2001 ["Puss in Boots"].

Uther, Hans-Jörg. *The Types of International Folktales: A Classification and Bibliography Based on the System of Antti Aarne and Stith Thompson*. Revised and expanded edition. Helsinki: Folklore Fellows Communications, 2011.

CINDY M. DORAN is a clinical pharmacist at a rural hospital. She is the author of *The Febrile Muse* (http://febrilemuse-infectious-disease.blogspot.com), a website that discusses the portrayal of infectious diseases in literature and the arts, with a special emphasis on science education and history.

14

MAN DISCOVERS A NEW LIFE-FORM AT A SOUTH AFRICAN TRUCK STOP

ROB DUNN

Like many biologists, the German biologist Oliver Zompro spends thousands of hours looking at specimens of dead animals. He found his first new species when he was twenty. By the age of thirty he had named dozens of wild new forms. While other people around him did crossword puzzles and drank lattes, he explored the world, one animal at a time.

Then, one day, things changed. He was looking through specimens when he found something more interesting than anything he had ever seen before. It was a fossil that looked like a cross between two different kinds of animals. It had the wrong mix of parts. It was—he would come to convince himself—a single individual of an entirely new order of beasts.

An order is one of the big categories of life, a big branch on evolution's tree. Animal species are named every day, but finding another new order would be equivalent to discovering bats if it had not been previously known they existed. Bats constitute their own order, as do primates, beetles, flies, and rodents.

It is easy to imagine that we have found all of them, living and dead. Yet the grass had parted for Zompro and revealed his treasure. He was not the first person to see it, but he was the first to recognize its significance and, he hoped, to give it a name.

But before Zompro went public with his find, he craved more specimens. He had found one specimen that other scientists had overlooked. It was at least theoretically possible that he might find others. And so he began to search, with zeal.

First, he visited the Natural History Museum in London. It is filled with dead animals and so a good place to begin. In the British museum he found many false leads. Then, remarkably, he found some real ones. There, in the collection, was a male very similar to the fossil he had seen in his native Germany, but with one key difference. The label attached to it indicated that it had been collected in Tanzania in 1955, alive. This new life-form might still be around, a living fossil.

Zompro had struck gold. Amazingly, with a little more digging, he then did it again. He found another specimen in the Museum für Naturkunde, Humboldt University in Berlin, this one a female from a 1909 collection in Namibia.

Zompro thought he had stumbled upon an entire evolutionary line that had survived the dinosaurs, survived the evolution of mammals, and now just maybe had survived several hundred thousand years of human troublemaking.

Quickly, Zompro, his advisor, and other colleagues wrote a paper on the new find in which they named the new order "Mantophasmatodea." Later the group would be given the common name "heel walkers," which makes one think of other beasts of lore—yeti, sasquatch, and the like. Each of the individuals Zompro had discovered was named as a separate species. History would soon decide whether the group was distinct enough to constitute its own order. In the meantime, Zompro and colleagues needed more specimens; they wanted to find these animals alive.

Zompro and colleagues decided go to Africa to look for more. Before they did, they needed to know where to look. Africa is big. This animal was, relatively speaking, small. The choice of sites was key, but little information existed on which to base the decision.

One of the specimens Zompro had found was from a relatively easy-to-reach site in Tanzania, but there was also a specimen from a far more isolated region of Namibia. In fact, the most recent specimens that turned up, one from 1991 and another from 2001, both came from the same place in Namibia, the Brandberg Massif. To the massif they would go. In such faraway places, they imagined, living fossils like the one they were looking for might survive.

To see Brandberg Massif in an aerial photo is unsettling. It is a circular plug of granite that rises straight out of the flat desert. Miles of desert extend from the massif in every direction. There are no surrounding hills, no trees, and no bumps of stubble. The massif, formed by the extrusion of lava, looks as though it were dropped onto the landscape. It is formidable, unique, and isolated, just the kind of place where one might find a yeti, or the ghost of an ancient animal.

The good news about the Brandberg Massif is that it is remote enough to preserve ancient and fragile life-forms, away from humans. The bad news is that it is remote enough to be really difficult to get to.

Zompro and colleagues decided that they would need to be flown in by helicopter to the massif. And so, in January 2002, after much planning and many signs that the expedition would never come to fruition, seven scientists were dropped off with great quantities of gear, cameras, collection devices, and food. They even brought porters. The venture had gone from a cheap collecting trip to an expensive "expedition," funded by Conservation International.

The idea was to hike from the landing site to a potential habitat, then hike out. Viewing photos from the trip, one can sense the excitement of the scientists. Perhaps there should also have been fear. They had spent a great deal of money. To return with no specimens would be tremendously disappointing.

Once the helicopter had left, the team began to search. They poked, prodded, chased, ran, and generally did everything they could to look everywhere a small, rare animal might hide. They looked in holes. But the truth was that while there were miles of desert, there were not really that many places to look. After a whole day nothing had been found, not a single clue. Then things changed.

Someone turned a leaf, and lo and behold, under it was a single beast. It hung there as though it had been waiting for centuries. Soon there were others. By the end of a week, thirty mantophasmatodes had been collected, observed, and fawned over. No one mentioned the heat. No one complained about anything. A few of these serious scientists began, uncontrollably, to smile.

The mission was not half over and already it was a success. In the meantime, the scientists now had to walk off the massif. Walking off had seemed at first a good idea. But it soon became clear that not everyone could carry all of his or her gear. People began to complain about the heat. They began to complain about the weight of their gear. Some contemplated turning back. One—Eugene Marais from the National Museum of Namibia—broke his ankle. Then, while slowly hopping down the hill on one leg, he grabbed a small tree for support and was stung by a scorpion resting on the tree. The biologists cursed one another. But no one cursed the mantophasmatodes, the tiny, living animals that they held aloft like kings.

Not long after the expedition, Zompro's first article about the new order was published in the journal *Science*. Quickly, the story appeared in magazines and newspapers around the world. Headlines proclaimed "Fossil Insect Found Alive." Zompro had not yet finished his Ph.D., but already major newspapers in a dozen countries had interviewed him.

What was more, time and analyses would prove the Mantophasmatodea to be just as unique as Zompro had initially believed, a new world order, or at the very least a suborder. And so he would have been justified had he imagined that other scientists would say things, admiringly, like "I can't believe you discovered this strange new animal!" Some did. But he also heard something different.

Mike Picker, a professor at the University of Cape Town, saw photos of the mantophasmatodes in a magazine. To him they did not look new and strange, they looked old and familiar. Picker recognized the mantophasmatodes as animals he had known of for years. He knew them from habitats all around Cape Town. Picker had looked straight at mantophasmatodes but had not realized they were something beyond another unnamed species.

Mantophasmatodes look, inescapably, larval (they lack wings, for example, and have no ocelli), and so Picker, like others, mistook them for immature versions of some known creature, perhaps some weird kind of cricket. When more than three-quarters of all species of animals are not yet named, it is hard to know which ones to get excited about finding. Picker went through his collections looking for specimens of mantophasmatodes. Within weeks, he had found twenty-nine individual mantophasmatodes. Thirteen living species of Mantophasmatodea have now been named, placed in ten genera and three different families.

In other words, Zompro has done something more amazing than finding a rare new order of animals. He has discovered a common order of animals that everyone else had missed, a discovery in plain view.

Mantophasmatodes are not a faraway species confined to some remote hunk of rock. They are a whole suite of species, some of which live in places as mundane as backyards. They are also a kind of living extended metaphor for what lurks around us unnoticed all the time.

Most days we forget about the grandness of the living world. In our offices and busy lives we gloss over what remains to be discovered. Yet there are still more unnamed species than named species on Earth, some of them very near to where we live. Discoveries are possible everywhere. If you have any doubt about this statement, you need not go any further than the most recent episode in the story of the mantophasmatodes, the one involving Piotr Naskrecki and a truck stop.

Piotr Naskrecki was one of the scientists on the first expedition to the Brandberg Massif. He lugged camera gear. He hiked down the hill. He came home victorious with the rest of the crew. And then he also went back to South Africa for other collecting trips.

It was on one of these recent trips to South Africa that he stopped at a "filthy truck stop on the major highway, N1," heading north out of Cape Town. There, as he is wont to do everywhere, he looked around, this time with his friend Corey. While other people searched for the bathroom, Naskrecki looked to see if he could find any interesting insects. He did.

He searched in much the way that he had searched for mantophasmatodes up on the massif. At first he found nothing, and then he found a new species of katydid, and then another, and then six more, and then, amazingly, another new species of mantophasmatode. The mantophasmatode he discovered, a pregnant female, is a still-unnamed species of the genus *Sclerophasma*.

She is lovely and interesting, but she is also evidence that the Mantophasmatodea, the first new insect order named since the early 1900s, an order whose discovery in the field required a trip to the remote Brandberg Massif, could have been discovered at a truck stop.

What was required to discover the mantophasmatodes, whether the species on top of the massif or the species at the truck stop, was the realization that no one else knew what they were. Once that realization was made, discovering them was both easier and more interesting. Until then, the mantophasmatodes, like much of life, seemed (wrongly) likely to be known by some expert in a university somewhere. Yet they were not known, just as most of life is not known.

It was only recently that it was discovered that mice sing to each other. It was not so very long ago that it was discovered that clouds are filled with bacteria. What else remains to be known? Nearly everything.

So pay attention when you are walking through forests and backyards and, yes, even truck stops. Take notes. Take pictures and assume that you are the very first one to see everything that you see. The life around us is as foreign as the dark side of the Moon; we just forget. You may find a new form of life, and you will certainly find new observations of behaviors.

But be forewarned. As Naskrecki can confirm, if you go around truck stops with a headlamp, vials, and ethanol at night, you *might* discover a new life-form, but you *will* have some explaining to do. So practice saying, in whatever language is appropriate, "Officer, the vials are for insects. I am trying to make a discovery." No one said it was easy being an explorer.

ROB DUNN is a science writer and biogeographer in the Department of Biology at North Carolina State University. His latest book, *The Wild Life of Our Bodies*, explores how changes in our interactions with other species, be they forehead mites or tigers, have affected our health and well-being. Rob lives in Raleigh, North Carolina, with his wife, two children, and more than two forehead mites. Find him at www.robrdunn.com.

15

MOON ARTS: FALLEN ASTRONAUT

CLAIRE L. EVANS

The Moon is a rock.

But it's also Selene, Artemis, Diana, Isis, the lunar deities; an eldritch clock by which we measure our growth and fertility; home of an old man in the West and a rabbit in the East; the site of countless imaginary voyages; a long-believed trigger of lunacy (luna . . . see?). It's another world, close enough to ours to peer down at us; to it we compose sonatas. It can be blue, made of cheese, a harvest moon; we've long fantasized about its dark side, perhaps dotted with black monoliths or inhabited by flying men.

The Moon is a totem of great importance in all religions and traditions; in astrology, it stands for all those things that make the more scientifically minded among us develop facial tics: the unconscious, parapsychology, dreams, imagination, the emotional world, all that is shifting and ephemeral. According to *The Penguin Dictionary of Symbols*, as the light of the Moon is merely a reflection of the light of the Sun, "the Moon is the symbol of knowledge acquired through reflection, that is, theoretical or conceptual knowledge."

All of this is to say that while the Moon is a rock, it's also an idea.

And, as an idea, it appeals to artists. The Moon, however, remains beyond the reach of artists by virtue of what makes it interesting to them: namely, its moon-ness, a perfect storm of mystery, opacity, and unreachability.

So just how do you implement the Moon in your practice when it's 240,000 miles away? As an artist, how do you stake a claim somewhere inside of the patriotic military-industrial research bureaucracy that controls the purse strings, and thus access to our nearest celestial bodies? There doesn't seem to be a direct entry. If you're part of the original Moon Museum posse, you go in the back door, sneaking your work illicitly onto the heels of a lunar lander. If you're Belgian artist Paul Van Hoeydonck, you meet astronaut David Scott at a dinner party.

Van Hoeydonck is responsible for the only piece of art on the Moon, a tiny memorial sculpture called *Fallen Astronaut*. The piece is interesting for several reasons. For one, it presents us with a clear understanding of the kinds of technical limitations that Moon artists must work under. Limitations, of course, can be instrumental to an artist's practice—a broke Basquiat painted on window frames and cabinet doors—but space art's parameters border on the draconian.

In the design of the piece, Van Hoeydonck was restricted to materials that were both lightweight and sturdy, as well as being capable of

(Source: NASA)

withstanding extreme temperatures. Since it was to be a memorial to deceased astronauts, it couldn't be identifiably male or female, or of any ethnic group. The somewhat questionable result: what looks like a metal Lego lying facedown on a thermal anomaly near Mons Hadley.

Like the Moon Museum, *Fallen Astronaut* was an unofficial venture; the statuette was smuggled aboard the *Apollo 15* lunar module by the astronauts themselves—Scott and Jim Irwin—without the knowledge of NASA officials. Its "installation" was unorthodox: in laying down the sculpture and its accompanying plaque, Irwin and Scott performed a private ceremony on the lunar surface. "We just thought we'd recognize the guys that made the ultimate contribution," Scott later said. Notably, "the guys" included eight American and six Soviet astronauts, so it was a surprisingly apolitical act of solidarity in the midst of the Cold War.

Scott and Irwin were committed to the sanctity of their memorial; when Scott plopped the piece onto the lunar dust, Irwin covered the act with inane radio chatter to Mission Control, and they didn't announce the memorial until after their return to Earth. Even then, the astronauts kept Van Hoeydonck's name private, hoping to avoid any commercial exploitation of the piece.

Van Hoeydonck, undoubtedly hoping to further his career, later violated the unspoken sacredness of *Fallen Astronaut* by attempting, in 1972, to sell hundreds of signed replicas of the piece at $750 a pop. We'd all recoil in horror if Maya Lin tried the same thing with the Vietnam Veterans Memorial, but I'm almost tempted to give Van Hoeydonck a pass. After all, the *Fallen Astronaut* itself is just a totem, and a toylike one at that.

I see this story as something of an inversion of the usual artist-scientist dialectic.

Van Hoeydonck, here, was essentially an engineer. All he did was design a tin man to technical specifications. It was Scott and Irwin who made the visionary decision to perform an unnecessary act of beauty on the chunk of rock orbiting our own. It was the astronauts who snuck the statuette all the way to the Moon and secretly installed it. They understood that beyond being a rock, the Moon is an idea, and that actions performed on the Moon by human beings are instantly imbued with meaning, historical significance, and some kind of indefinable holiness.

Scott, Irwin, and NASA all balked at Van Hoeydonck's commercial enterprise, and the artist eventually retracted it, instead donating replicas of *Fallen Astronaut* to various museums and keeping the rest for himself, unmonetized.

While it's ordinarily the artists who defend the formal importance of ideas for their own sake, on *Apollo 15* it was, well, not the scientists—but the military-trained, engineer-pilot, non-artist astronauts who did. Which perhaps goes to show that the experience of space, the perspective-altering transcendence of the so-called Overview Effect, that euphoric feeling of universal interconnectedness experienced by astronauts during spaceflight, ultimately turns us all into poets.

CLAIRE L. EVANS is a writer and artist working in Los Angeles. In addition to performing as a multimedia musician in the band YACHT, she works as a science journalist and pens a blog, *Universe*, which addresses the synchronies among art, science, technology, and the cultural world at large. Find her at www.clairelevans.com.

16

SCIENCE METAPHORS: RESONANCE

ANN FINKBEINER

My mother was an old lady, she'd lived a good and useful life, and she died a year and ten days ago. I hadn't been keeping track of her death's anniversary but I didn't need to; I only had to figure out why I was walking around feeling, for no good reason, sad. One of my cousins wrote to me, "I'm sorry you are sad, Annie. Thursday my new couch was delivered. I cried as my old one left—apparently I had an undiscovered attachment to it." My cousin recently moved to a new house with a new love; the old couch was from her old life, years ago, when her husband had died.

Sad in early August? Crying over couches? Really? Science, as it often does, has a nice metaphor: resonance.

When I was a kid, I'd open the piano lid and tap a string and it would hum; and then like magic, all on its own, another string hummed, too. I thought the strings were malfunctioning and quit thinking about it. Obviously I was a young English major; a young physicist would have tapped some more strings and then looked it all up and found that the strings hummed because they were vibrating, oscillating, and that each string had its own native note, its natural frequency of oscillation, a resonant frequency. And if a second string hums at that particular frequency, then as if in sympathy the first string hums, too.

In fact, everything—atoms, molecules, crystals, desks, skyscrapers—has its own resonant frequency at which, depending on its mass and stiffness, it most wants to oscillate. So resonance is a response; it's when two things are linked in some way or somehow touching, and then one taps or vibrates or oscillates and the other responds at that same frequency.

If you're an army crossing a suspension bridge, for instance, you shouldn't march in cadence in case your cadence happens to match the bridge's resonant frequency and the bridge jumps in step with you. If you rock a rocking chair too fast or slow, the rocker rocks twitchily; rock at the rhythmic resonant frequency and the baby falls asleep. If you pump a swing too fast or slow, nothing much happens; pump at the resonant frequency, it's almost like flight. If you're in a building and the earth quakes at the building's resonant frequency, get out fast.

Resonance is actually more complicated than this. Resonances can be forced or driven, all the way to disaster. Pump the swing harder and harder at that right frequency and you risk breaking your neck. Sing loudly at a wineglass at the right frequency and it'll shatter. Google a video of the resonating Tacoma Narrows bridge, nicknamed Galloping Gertie, before it fell into the river. The genius inventor Nicola Tesla supposedly said that if he could figure out the resonant frequency of Earth, he could split it to pieces.

But disaster isn't part of this metaphor. The metaphor is simple resonance—like the sympathetic piano strings, like the resonance you notice when you first meet someone and know you're going to get along. We use the metaphor all the time: we're in tune, we're on the same wavelength, we're in synch. Death set up a resonance with early August, with a couch; another early August, a different couch, and there like magic is death.

ANN FINKBEINER is a freelance science writer based in Baltimore. She writes mostly about astronomy, runs a small graduate program in science writing at Johns Hopkins University, and blogs at the jewel-like *The Last Word on Nothing*.

17

BOMBARDIER BEETLES, BEE PURPLE, AND THE SIRENS OF THE NIGHT

JENNIFER FRAZER

Thomas Eisner once had a pet thrush named Sybil who rejected only five insects out of the hundreds the entomologist offered her. They were all beetles. And one of them was a firefly.

For any other bird owner, this observation would have simply limited their pet's meal options. But this was Thomas Eisner, one of the great entomologists and chemical ecologists of the twentieth century. To him, it was a tantalizing clue, and he decided to find out what made the fireflies have all the thrush plate appeal of haggis. What he stumbled onto was one of the great new natural history stories of the twentieth century—and the latest in a string of Eisner's greatest hits.

I know this because in the fall of 1998 I was a student in BioNB 221—Introduction to Animal Behavior—at Cornell University. Eisner taught the last six or so lectures, which I still have preserved in my notes. I did not know at the time that Eisner was one of the great biologists of the twentieth century. I found this out in later years, when his discoveries were featured in many an article in *The New York Times*.

What I did know was that I could not take my eyes off the screen while he was lecturing. I'm a fan of a good natural history story. Eisner—who was once E. O. Wilson's college roommate—was overflowing with them, in many cases because he'd figured them out himself.

Sadly, Eisner died on March 25, 2011. Natalie Angier, whose daughter was lucky enough to inherit the contents of Eisner's old burlap field bag, wrote a fine remembrance in *The New York Times*. She wrote about his life. I want to share with you a few of the natural wonders I learned from him, sitting rapt in the darkened Uris Hall auditorium.

(Courtesy of Kevin Collins)

This Means War

Eisner's specialty was the world of chemical warfare among plants and insects. Insects produce, steal, and reuse chemicals from plants and each other constantly. Millipedes can deploy hydrogen cyanide, whip scorpions acetic acid, and ants formic acid, but for Eisner, the poster child for entomological chemical defense was the bombardier beetle. "If you live on the ground," he said, "you must either take flight quickly or defend yourself instantly." The bombardier beetle went with option B.

The beetle takes chemicals called hydroquinones, mixes them with hydrogen peroxide and catalytic enzymes (peroxidase and catalase) in a reaction chamber in its hinder, and uses the resulting explosive formation of benzoquinones and heat to persuade frogs, ants, and spiders that their best meal options lie elsewhere.

Using grainy films he had shot himself, Eisner showed us how beetles touched with probes could deploy a vicious defense with pinpoint accuracy in nearly any direction. He suspended the beetles over pH paper, so the 100°C benzoquinones they released would reveal their precision firepower.

Don't Feed on Me

Plants, too, load up on poison in hopes of warding off the hungry crowd. Nettle spines are filled with irritating chemicals, as are the latex canals or resin canals of flowering plants and conifers, respectively. Some plants store poisonous chemicals in their tissues, such as caffeine or nicotine, which, in spite of their uplifting effects on humans, are actually insecticides.

But some insects have picked up on this gig and begun using it to their advantage. Sawflies slice into the resin canals of pines and steal the sticky sap, storing it in special sacs for defense against ants. Monarch butterfly caterpillars sequester milkweed toxins from their food, rendering themselves distasteful to predators for the rest of their lives. Assassin bugs coat their eggs with the noxious secretions from camphorweed. Their young then reuse the chemicals for defense and to catch prey. We do this, too, Eisner pointed out, by stealing the defense chemicals from fungi and bacteria. We call them antibiotics.

Eisner told us of plant chemicals stolen and presented as nuptial gifts among moths, where females choose males whose flirting, aromatic antennae tell them they have stored the most alkaloid derivatives. That implies both that the male is fittest and that he has the most to give to the pair's offspring. For if the female mates, the male will transfer not only his sperm, but his alkaloid collection, which the female will carefully store with her eggs for the use of her young. Other moths do the same with salts they siphon from puddles.

He told us of the eavesdropping of kairomones. Unlike pheromones used for intraspecies communication (as in moths), or allomones, which benefit the emitter in an interspecies pair (like the benzoquinones of the bombardiers or the stink of skunks), kairomones benefit the receiver and betray the emitter.

Think of the carbon dioxide that gives you away to mosquitoes; any scent, really, that betrays prey to predators can qualify. Eisner called that telltale whiff a "chemical gestalt," the effect of "inevitable chemical leakage." But the tables can also be turned. Predatory rotifers called *Asplanchna* unwittingly emit chemicals that alert prey rotifers called *Brachionus* to grow defensive spikes.

One of my favorite Eisner stories was about true bugs entomologists were attempting to rear in petri dishes on damp paper towels. The bugs' development was stalling, however; they could not be coaxed to adulthood. The scientists were baffled—until someone noticed the paper towels were made from balsam fir, a tree that emits allomones to stunt insect development. This chemical was surviving the paper-making process and continuing to thwart the trees' insect enemies—even in death.

You may have heard of parasitoid wasps—the *Alien*-style predators of spiders, caterpillars, and other insects that lay their eggs in their prey, where the wasp maggots hatch and proceed to devour their still-living hosts' organs before finally using their hosts' spent husks as pupae from which young wasps emerge. Perhaps you did not know that some plants injured by caterpillars or aphids call out chemically to parasitoids to defend them. But the story gets better; the immune system of the host insect in some cases is destroyed by viruses injected by the parasitoid wasps along with their eggs. "And [I underlined this in my notes] the viruses have also been incorporated into the wasp genome." To which I further wrote, "Whoa."

Eisner told us how mammals, too, use pheromones. Babies can distinguish their own mother's milk from others', he said, and the scent of male armpits can regularize erratic female ovulation. In mice, the scent of strange male urine blocks implantation of fertilized eggs in female mice; the effect, and the reason, may be akin to mares aborting fetuses to save themselves from investing energy in foals likely to be killed by rival stallions anyway. This could explain the spectacularly high miscarriage rate (around 30 percent) in mares who are trucked out to mate with top stallions but housed while pregnant with other males. That this is likely to have not one whit of effect on the way breeders practice horse husbandry is testament to the often hidebound thinking of humans.

The Bee-Letters of Flowers

On top of doing all this research into chemical cross fire, Eisner also dabbled in the world of light and visual communication. Those who have studied physics know the electromagnetic spectrum of which light is a part is a vast array of energy. Earth's atmosphere filters much that arrives, and most of what makes it through falls in the 320 to 2300 nm range. What we perceive as visible light falls in the 380 to 750 nm range. But that leaves a large part of the spectrum invisible to us. What if other animals could see different colors or different parts of the electromagnetic spectrum? As it turns out, they do.

We cannot see ultraviolet. But, through experiments worked out by a whole host of Germans, we know bees do. Conversely, bees cannot see red. Their vision lies in the 340–650 nm range. Blue, red, and green are the human primary colors. But the bee primary colors are yellow, blue, and ultraviolet. That implies there is a spectrum of colors they see that we cannot.

And those colors needed names. Yellow+blue we can see along with bees—we call that blue-green. But what about blue+ultraviolet? That was dubbed "bee blue." Yellow+ultraviolet? "Bee purple." And, as it turns out, flowers are adorned in these shades, invisible to us but brilliantly displayed for bees. Flowers probably first used UV-absorbing pigments as sunscreen, Eisner said, and only later turned to them to decorate their petals. Now, bee blue and bee purple form pollen guides, often invisibly (to us) flecking the tips of flowers and leaving a yellow disc in the center as a bee bull's-eye.

The inability of bees to see red means that pink or red flowers are almost never pollinated by bees. On the contrary, only butterflies and hummingbirds—which are not red-blind—are attracted to red flowers.

Which brings us back to what is likely Eisner's most famous experiments in light communication: the Tale of the Fireflies *Photinus* and *Photuris*. Following up on the expectorated clues provided by Sybil, Eisner extracted chemicals from fireflies of that species with various solvents. He discovered that the firefly she spat out—*Photinus*—contained a steroid called lucibufagin. When fireflies are caught, they "bleed" hemolymph full of this chemical. Spiders who catch them and taste this let them go. They even release fruit flies merely painted with the chemical, the scientists discovered.

Eisner found *Photinus* was chock-full of the chemical right from the start of the season. A larger firefly, *Photuris*, also contained this chemical. But only the females. And only later in the season. He began to glimpse the truth of a dark story.

Male fireflies searching for females make a species-specific pattern of flashes. Females respond with a single blink, but with a species-specific time delay from the male call. *Photuris*, coveting the chemicals of *Photinus*, imitates a *Photinus* female's response time. When the male lands thinking he is about to get lucky, he gets eaten instead, and the female accumulates the chemical that allows her to escape predation by spiders and, yes, thrushes.

How could one man do and learn so much? Perhaps because he never let the Lab get in the way of Life. This passage from Angier's piece, in particular, explains why I love Eisner—and to a large degree why being a modern biologist was not for me.

> Ian Baldwin, a professor of molecular ecology at the Max Planck Institute for Chemical Ecology in Germany, who studied with Dr. Eisner in the 1980s, said of his mentor: "He articulated the value of natural history discovery in a time of natural history myopia. We train biologists today who can't identify more than four species, who only know how to do digital biology, but the world of analog biology is the world we live in. Tom was a visionary for nonmodel systems. He created narratives around everything he did."
>
> In today's "shiny polished science world, he was proof that there is no experience that can substitute for being out in nature," said Dr. [May] Berenbaum. "It's classy, not low-rent, to stay grounded in biological reality."

Thank you for the stories, Dr. Eisner, wherever you are.

REFERENCES

Angier, Natalie. "Paths of Discovery, Lighted by a Bug Man's Insights." *New York Times*. April 4, 2011. www.nytimes.com/2011/04/05/science/05angier.html.

JENNIFER FRAZER is an AAAS Science Journalism Award–winning science writer who lives in Colorado. She has degrees in biology, plant pathology/mycology, and science writing from Cornell University and MIT, and has spent many happy hours studying life in situ. Her blog *The Artful Amoeba* is now part of the Scientific American Blog Network; you can find it at http://blogs.scientificamerican.com/artful-amoeba/.

18

MPEMBA'S BAFFLING DISCOVERY

GREG GBUR

"My name is Erasto B Mpemba, and I am going to tell you about my discovery, which was due to misusing a refrigerator."

With those words, Tanzanian student Erasto Mpemba entered scientific history, and also sparked a scientific mystery and controversy that remains ongoing today, some forty years later.

The phenomenon Mpemba found is now known as the Mpemba effect. It is the very counterintuitive idea that, under certain circumstances, a quantity of very hot, or even boiling, liquid can freeze faster than an equal quantity of cold liquid.

How is this possible? The remarkable thing is that nobody really knows, even though the first observations were reported to the scientific community in 1969. The story of the discovery, and the consequent mystery, is worth a bit of exploration—and the Mpemba effect carries numerous important lessons about the nature and method of scientific discovery.

Mpemba made his accidental discovery in Tanzania in 1963, when he was only thirteen years old and in secondary school. In spite of widespread disdain from his classmates, he surreptitiously continued experiments on the phenomenon until he had the good fortune in high school to interact with Professor Denis Osborne of the University of Dar es Salaam. Osborne was intrigued, carried out his own experiments, and in 1969 the two published a paper in the journal *Physics Education*.[1]

This article is one of the most remarkable in all of the history of physics. Aside from its title—"Cool?"—it is also unusual in being presented in two parts: in the first half Mpemba gives a first-person account of his

Erasto Mpemba in 2011 (Courtesy of TEDxDar 2011)

discovery, and then Osborne picks up the story and describes the follow-up experiments in the second half.

Mpemba's account begins with the misuse of a refrigerator. A common activity among the boys of his age was making ice cream from scratch. They would boil milk and mix it with sugar, and then put the combination in the freezer section of the refrigerator. The combination was always allowed to cool to room temperature first, due to a fear that putting boiling mixtures in the refrigerator would strain it and eventually cause it to break. Limited space in the freezer, however, led to a race among students:

> One day after buying milk from the local women, I started boiling it. Another boy, who had bought some milk for making ice-cream, ran to the refrigerator when he saw me boiling up milk and quickly mixed his milk with sugar and poured it into the

> ice-tray without boiling it, so that he may not miss his chance. Knowing that if I waited for the boiled milk to cool before placing it in the refrigerator I would lose the last available ice-tray, I decided to risk ruin to the refrigerator on that day by putting hot milk into it. The other boy and I went back an hour and a half later and found that my tray of milk had frozen into ice-cream while his was still only a thick liquid, not yet frozen.
>
> I asked my physics teacher why it happened like that, with the milk that was hot freezing first, and the answer he gave me was that "You were confused, that cannot happen." Then I believed his answer.

Here we have the beginnings of a classic story of science—an accidental discovery, scoffed at by the "establishment scientists."

Mpemba might have given up at that point, but he encountered a friend who sold ice cream for a living, and that friend happened to mention that many vendors would use boiling water to make their ice cream. It was already common knowledge among them, apparently, that a boiling mixture could freeze more quickly. When Mpemba reached high school, he decided to ask his teacher why hot milk would freeze faster than cold. The teacher's response was not very encouraging:

> "The answer I can give is that you were confused." I kept on arguing, and the final answer he gave me was "Well, all I can say is that is Mpemba's physics and not the universal physics." From then onwards if I failed in a problem by making a mistake in looking up the logarithms this teacher used to say "That is Mpemba's mathematics."
>
> And the whole class adopted this, and anytime I did something wrong they used to say to me "That is Mpemba's . . .", whatever the thing was.

Here the high school teacher failed miserably—ridiculing a student is pretty much the worst thing one can do in a science classroom. Fortunately, Mpemba was not deterred; when he found a biology lab open one afternoon, he filled two beakers, one with boiling water and one with tap water, and froze them in the laboratory refrigerator. When he returned an hour later, he found there was more ice in the (formerly) boiling water. With this suggestive result to his experiment, he resolved to try it again when the opportunity arose.

Before he had this chance, however, Professor Osborne came to lecture on physics, giving Mpemba an opportunity to talk to an expert on the subject. He repeated his question, as he had to his secondary school teacher and his high school teacher, but this time he got a different reaction:

> He first smiled and asked me to repeat the question. After I repeated it he said: "Is it true, have you done it?" I said: "Yes." Then he said: "I do not know, but I promise to try this experiment when I am back in Dar es Salaam." Next day my classmates in form six were saying to me that I had shamed them by asking that question and that my aim was to ask a question which Dr Osborne would not be able to answer. Some said to me: "But Mpemba did you understand your chapter on Newton's law of cooling?" I told them: "Theory differs from practical."

There are many remarkable points in this short passage. First of all, we see an admirable open-mindedness of Professor Osborne in his dealings with Mpemba, and that open-mindedness would quickly benefit them both. Conversely, we see a dangerous "groupthink" among Mpemba's classmates regarding science, in which they are genuinely offended by Mpemba questioning the status quo. Mpemba shows great wisdom in his answer: "Theory differs from practical." This is an important point for anyone studying physics: we like to create simplified models to explain nature, but those models often lose real-world aspects in the process of stripping them down.

Mpemba actually continued his experiments in a kitchen refrigerator, with the permission of kitchen staff, and convinced his classmates and the headmaster of his school of the accuracy of his findings.

At Dar es Salaam, Osborne was true to his word and looked into the phenomenon himself. As he notes in the continuation of the paper,

> It seemed an unlikely happening, but the student insisted that he was sure of the facts. I confess that I thought he was mistaken but fortunately remembered the need to encourage students to develop questioning and critical attitudes. No question should

> be ridiculed. In this case there was an added reason for caution, for everyday events are seldom as simple as they seem and it is dangerous to pass a superficial judgment on what can and cannot be. I said that the facts as they were given surprised me because they appeared to contradict the physics I knew. But I added that it was possible that the rate of cooling might be affected by some factor I had not considered.

Osborne sets a great example for all physics educators. It can be difficult at times, but "No question should be ridiculed" would be a great part of a "Hippocratic oath" for teachers.

So what did Osborne's research show? He placed a 100 cm^3 beaker filled with 70 cm^3 of water on a sheet of insulating foam in a freezer and timed how long it took for the water to freeze. For temperatures up to 20°C, the time was roughly proportional to the temperature above freezing, up to a maximum of 100 minutes at 20°C. For higher temperatures, however, the time dropped dramatically, down to 40 minutes for 80°C water. Amusingly, a technician who was tasked to perform the experiments first reported back to Osborne that he in fact did find that hot water freezes before cold, but that he would repeat the experiment until he got the proper results.

Later experiments on the effect have been far less conclusive than Mpemba and Osborne's. Some have seen similar results, while others have observed no effect at all! The appearance of the Mpemba effect apparently depends very strongly on the specific experimental circumstances, and is much harder to reproduce than the original paper would imply. A number of physicists seem skeptical that the effect truly exists.

With this in mind, however, it is worth noting that Mpemba's observation had, in fact, been noted by others over the course of the past two thousand years. Around 350 B.C.E., Aristotle, in giving an explanation of hail in his book *Meteorology*, noted:

> The fact that the water has previously been warmed contributes to its freezing quickly; for so it cools sooner. (Hence many people, when they want to cool water quickly, begin by putting it in the sun. So the inhabitants of Pontus when they encamp on the

> ice to fish (they cut a hole in the ice and then fish) pour warm water round their rods that it may freeze the quicker; for they use the ice like lead to fix the rods.) Now it is in hot countries and seasons that the water which forms soon grows warm.[2]

Others who argued for the existence of a Mpemba effect, under some circumstances, include Roger Bacon in the thirteenth century, and Francis Bacon and René Descartes in the seventeenth century. It was Mpemba and Osborne, however, who brought it to the attention of modern science.

Before delving into some of the explanations for the effect, it is important to explain why the natural objection to its existence is not necessarily applicable. A natural argument is as follows: "Suppose we have two equal glasses of water, one at 100°C and one at 30°C. In order for the water at 100°C to freeze, it must first cool to 30°C. The time it takes to freeze is therefore the amount of time it takes to go from 100 to 30 plus the time it takes to go from 30 to zero. It therefore must take longer to freeze than the 30°C water."

The flaw in this argument is assuming that a body of water is characterized by only one parameter: its temperature. In general, there are other characteristics of the water that could be changed due to its heating: for instance, the amount of gas in solution; the presence of other solutes; the presence of convection currents; gradients of the distribution of temperature in the container. Any, all, or none of these may in fact be the culprit, but it is important to realize that a body of water not in thermal equilibrium may have its behavior characterized by a number of different properties.

A number of hypotheses have been proposed as explanations, some based on the factors mentioned:

1. *Convective heat transfer.* When a liquid is heated, it can form convection currents that rapidly bring the hot liquid to the surface, where the heat is lost by evaporation. Osborne noted that this convection will keep the top of the liquid hotter than the bottom, even when the mean temperature matches that of an initially cold liquid that doesn't possess this convection cooling. This results in a faster rate of cooling that could, under the right circumstances, result in Mpemba's observation.

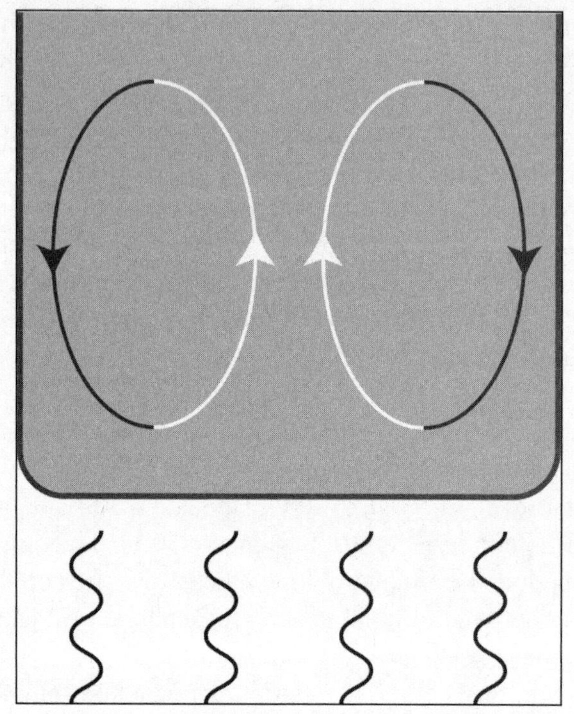

Illustration of convection currents. Hot liquid is continuously being circulated to the surface. (Greg Gbur)

2. *Evaporation.* A boiling or very hot liquid will lose some of its mass due to evaporation. With a lower mass, it will cool faster, possibly giving a "boost" to the Mpemba effect. Osborne noted that evaporation alone could not completely account for the rate of cooling of the hot liquid.

3. *Degassing.* In a series of experiments in 1988, a Polish research group reproduced the Mpemba effect and noted that the effect depended strongly on the amount of gas dissolved in the water.[3] When the water was purged of air and carbon dioxide, the time to freeze became proportional to the starting temperature. The researchers suggested that the presence of gas was substantially slowing the rate of cooling, and that the heated water was purged of it.

4. *Supercooling.* In 1995, German scientist David Auerbach proposed that the Mpemba effect could be explained by supercooling, and performed experiments to back up the assertion.[4] Liquids begin to crys-

tallize into a solid at the freezing point with the help of impurities around which the crystals can nucleate. In the absence of such impurities, the liquid can be cooled below its normal freezing point while remaining a liquid—it is "supercooled." Auerbach suggests that the cold water will supercool to a lower temperature than the hot water, thus giving the hot water an "edge" in the race to freeze.

However, the reason that hot water has a higher supercooling temperature is unclear, and possibly related to the earlier-noted effects.

5. *Distribution of solutes.* In 2009, an interesting explanation was proposed by J. I. Katz, who suggested that solutes present in the cold water slow the freezing process, as suggested earlier, but also that those solutes get driven from the freezing water into the as-yet-unfrozen water, slowing the process further.[5]

Other suggestions have been made, and other experiments have been done,[6] though no consensus seems to have been reached regarding the origins and generality of the effect. The confusion about the Mpemba effect is curiously reminiscent of the controversy surrounding the "Archimedes death ray," a focused beam of sunlight that Greeks around 200 B.C.E. supposedly used to ignite Roman ships[7]—the answer depends on the question being asked. Does hot water always freeze faster than cold? Almost certainly not! Are there some circumstances under which hot water freezes faster than cold? It seems likely, though nobody is sure exactly what those circumstances are. Does a negative result disprove the Mpemba effect, or was the experiment just done under the wrong circumstances? Nobody knows for certain.

The controversy is no longer of concern to Erasto Mpemba, however, who did not become a physicist himself but ended up studying at the College of African Wildlife Management. With his education, he eventually became the principal game officer for the Tanzanian Ministry of Natural Resources and Tourism, working on wildlife management and conservation.

Mpemba's discovery is a wonderful illustration of many important lessons in science: the significance of experiment over theory, the danger of clinging to preconceived notions, the difficulty in evaluating even seemingly simple real-world physics problems, and the importance of perseverance in the face of unreasoning denial. As far as I know, the

Mpemba effect is the only physical phenomenon named for an African researcher. It is a wonderful and frustrating hint of how much intellectual potential lies untapped on the continent.

NOTES

1. Erasto B. Mpemba and Denis G. Osborne. "Cool?" *Physics Education* 4 (1969): 172–75.
2. Aristotle, *Meteorology*, in *The Complete Works of Aristotle: The Revised Oxford Translation*, vol. 1, ed. Jonathan Barnes. (Princeton, N.J.: Princeton University Press, 1984).
3. B. Wojciechowski, I. Owczarek, and G. Bednarz, "Freezing of Aqueous Solutions Containing Gases," *Crystal Research and Technology* 23, no. 7 (1988): 843–48.
4. David Auerbach, "Supercooling and the Mpemba Effect: When Hot Water Freezes Quicker than Cold," *American Journal of Physics* 63, no. 10 (1995): 882–85.
5. Jonathan I. Katz, "When Hot Water Freezes Before Cold," *American Journal of Physics* 77, no. 1 (2009): 27–29.
6. S. Esposito, R. De Risi, and L. Somma, "Mpemba Effect and Phase Transitions in the Adiabatic Cooling of Water Before Freezing," *Physica A: Statistical Mechanics and Its Applications* 387, no. 4 (2008): 757–63.
7. drskyskull [Greg Gbur], "Mythbusters Were Scooped—by 130 Years! (Archimedes Death Ray)," *Skulls in the Stars* (blog), February 7, 2010. http://skullsinthestars.com/2010/02/07/mythbusters-were-scooped-by-130-years-archimedes-death-ray/.

GREG GBUR is an associate professor of physics and optical science at the University of North Carolina at Charlotte, specializing in theoretical optics. He has been blogging at *Skulls in the Stars* since 2007; he is also cofounder of the history of science blog carnival *The Giant's Shoulders*: http://ontheshouldersofgiants.wordpress.com/. Early in 2011 he published a textbook, *Mathematical Methods for Optical Physics and Engineering*, and his writing has previously appeared in *The Open Laboratory 2010*.

19

ROMEO: A LONE WOLF'S TRAGEDY IN THREE ACTS

KIMBERLY GERSON

Act I: Into the covert of the wood

Many a morning hath he there been seen,
With tears augmenting the fresh morning dew.
Adding to clouds more clouds with his deep sighs . . .

—*Romeo and Juliet*

In April 2003 a female Alexander Archipelago wolf (*Canis lupus ligoni*) was hit by a taxi near the Mendenhall Glacier Visitor Center in Juneau, Alaska. Not killed instantly, the black wolf was euthanized by police. She had been pregnant with four pups, probably due to be born within the next few weeks. Alaska State Troopers were called to retrieve the body.

The following winter, visitors and nearby residents began to hear the soulful cries of a single wolf howling across the lake.

Then, on a winter day in 2003, local resident and wildlife photographer Nick Jans was out skiing and spotted wolf tracks. The tracks led him to a young black wolf that was acting "goofy, gangly and clumsy like a teenager." The wolf took an interest in Jans's dogs and soon became a regular visitor to the area.

In the winter, the Mendenhall Valley and the frozen lakes within it are popular destinations for campers, skiers, skaters, and snowshoers. As the 2003–2004 winter season went on, more visitors and local residents reported seeing the lone wolf, presumably the dead female's mate (or maybe her son). With each report he seemed more bold, more interested in them—and in particular, in their dogs. Jans told reporters: "He developed a huge crush on our female lab, Dakotah, and that's how he got his name. He would hang around our back door and sometimes be waiting in our yard. My wife, Sherrie, said, 'There's that Romeo wolf again.'"

The name stuck.

Act II: The bones of buried ancestors

I have night's cloak to hide me from their sight;
And but thou love me, let them find me here:
My life were better ended by their hate,
Than death prorogued, wanting of thy love.
—Romeo to Juliet

The Alexander Archipelago wolf is a rare subspecies of the gray wolf unique to the islands of southeastern Alaska. With Romeo's arrival, people began looking forward to encounters and photographic opportunities with this normally reclusive species. And with a little encouragement, Romeo became bold, approaching people and inviting their dogs to play.

In 2007, an e-mail with photos of the wolf picking up a pug and carrying it the way he might carry a dead rabbit came to the attention of wildlife officials. The pug was released unharmed (some observers say Romeo was just playing), but there were also troubling reports of people touching the wolf. These encounters confirmed that Romeo was becoming far too accustomed to humans and pets. It would be only a matter of time until someone got hurt. And sure enough, it was not long before the wolf was accused of killing a small dog.

Not only that, but there was a growing concern for Romeo himself. Pet dogs could transmit diseases to the wolf and put other wolf populations at risk. Or, if Romeo became aggressive to humans, he would risk being killed as a dangerous animal.

Alaska Department of Fish and Game authorities toyed with different ideas on how to manage the situation. The wolf could be trapped and moved. That would be a straightforward option. But Alaskans choose to live among their wildlife. Black bears are ubiquitous in and around many Alaskan cities. Grizzly bears fish for salmon alongside local fishermen. Moose walk the streets, and most folks know what it's like to lie in bed and hear wolves howling not far from their windows.

And besides, the residents of Juneau loved Romeo. "Friends of Romeo" had been formed in 2006 to "speak for Romeo," defending him against claims that he had killed pet dogs, and arguing for his right to remain in his homeland. Sending him away would not be an option.

So authorities settled on trying to educate people on how to discourage the engaging wolf. They posted signs at the visitor center warning hikers to keep their dogs leashed and not to let them play with their wild cousin. They reminded visitors that fines could be levied on those who fed or touched the animal. "There's the danger of loving this wolf to death," warned Juneau District Ranger Pete Griffin. "We have to remember that it is a wild animal. For it to continue to survive it has to remain a wild animal."

But the wolf, now habituated to his human and domestic dog "pack," persisted, and attempts to change people's behavior in dealing with Romeo were spectacularly unsuccessful. Local residents allowed Romeo to accompany them on daily walks, sometimes for hours at a time. The wolf appeared in the yards of local dogs he was fond of and followed skiers back to their cars. He played with tennis balls and frolicked freely with pets while their owners skated on the frozen lake. And analysis of his scat showed that along with small animals, fish, and the occasional deer, he was also eating dog food.

Act III: Feasting with mine enemy

Where the devil should this Romeo be?
Came he not home to-night?

January 2010 brought ominous news: Romeo had not been seen all winter. Harry Robinson, a local resident who walked with the wolf daily, reported that he hadn't seen him since September 18, 2009. Speculation ran the gamut: Maybe Romeo was dead, killed in an accident or by a pack of wolves. Maybe he simply died of old age. Or maybe he was alive and well. Perhaps he had finally found a female of his own kind and returned to a more normal wolflike existence. Everyone hoped for the latter, though chances were remote.

John Hyde, a wildlife photographer who spent the better part of seven years with Romeo, reported that Romeo showed little interest in other wolves, not answering the howls of distant packs. (Hyde had gained the animal's trust to such an extent that he was rewarded with some of the most beautiful and profound images of the creature ever published.) Biologist Steve Lewis was convinced that Romeo had probably died of natural causes. "My guess is that he's probably dead . . . he just died from being a wolf," he said.

Romeo's disappearance triggered a community response. Reward money was put up and missing posters were posted all over Juneau. Friends of Romeo began investigating, and in May 2010 an arrest was made: Park Myers of Juneau and Jeff Peacock of Lebanon, Pennsylvania, were charged with a number of hunting violations, including an illegal wolf kill. The wolf, while not confirmed as Romeo, was an adult male taken from the Mendenhall Lake region on September 22, 2009—four days after Romeo went missing. And it was black.

Photographs confiscated from Myers and Peacock show a lone black wolf confidently approaching vehicles in a parking lot, later the two men posing with the carcass of a black wolf, and an image of a *Juneau Empire* article entitled "Romeo, Where Art Thou?" DNA analysis of the wolf was never performed, but Harry Robinson examined the photos and identified the wolf by a recognizable scar. "That is definitely Romeo," he pronounced.

The hunters were found guilty of numerous hunting violations, and the penalties handed down were typical of those for first-time violators—fines of several thousand dollars, suspended six-month jail sentences, and loss of their hunting licenses for three years. As much as he was loved by the Juneau community, Romeo was still just a wolf, with no special rights or status beyond those of the rest of his kind, or of the bears that were also illegally taken by the convicted hunters.

Seal up the mouth of outrage for a while,
Till we can clear these ambiguities,

And know their spring, their head, their true descent;
And then will I be general of your woes,
And lead you even to death: meantime forbear,
And let mischance be slave to patience.
Bring forth the parties of suspicion.

Like his namesake, Romeo crossed a dangerous social boundary and for that he paid with his life. To be sure, those men wielding the guns hold the bulk of responsibility. But the folks who cared so much for Romeo also bear some of that burden.

A lone wolf is no wolf at all. Romeo needed a pack. With no other wolves to choose from, he tried to build a foster pack of pet dogs, and by extension, their owners. These were the people who failed Romeo. Instead of rebuffing him, they encouraged him.

Many of them should have known better—wildlife photographers and nature writers; the residents of Mendenhall Valley, who presumably should have more sense than tourists when it comes to feeding and encouraging wildlife. And most important, wildlife authorities and biologists who are entrusted to look out for the best interests of animals like Romeo.

Because he had been conditioned to trust humans, Romeo was a sitting duck for two men looking for an easy kill. But rather than protect him from that inevitability, officials caved to community pressure.

"People have a great sense of community pride regarding the wolf, and that should continue as long as people act responsibly around it," said District Ranger Pete Griffin in 2007. "It is a problem, and we don't want anybody to even have to worry about getting a citation," said Alaska Bureau of Wildlife Enforcement trooper Lieutenant Todd Sharp.

But these officials were responsible for drawing that firm line in the sand that separates wild animals from overzealous public affection, and they didn't do it. And why, instead, didn't they try to find a way to give Romeo what he really needed—a wolf pack?

By loving him to death, the residents of Juneau didn't just lose a wolf. Romeo was among the last of his kind. There are fewer than 1,000 Alexander Archipelago wolves left in Southeast Alaska. When Romeo died, he took with him some of the most valuable wolf DNA in existence.

This rare and valuable genetic line could have helped preserve the species and perhaps even have enriched the genome of his ancestral bloodline, the gray wolf, *Canis lupus.*

In this imperiled species, the death of Romeo, a solitary wolf, brings us one step closer to the day when the only Alexander Archipelago wolf that locals and visitors will ever see is the one that was killed by the taxi in 2003—now stuffed and mounted in the Mendenhall Glacier Visitor Center—and named Juliet.

> Go hence, to have more talk of these sad things;
> Some shall be pardon'd, and some punished:
> For never was a story of more woe
> Than this of Juliet and her Romeo.
> *Exeunt.*

REFERENCES

Anchorage Daily News, via Associated Press. "Juneau Predator Catches and Releases Pet Pug." February 9, 2007. www.adn.com/2007/02/09/207773/juneau-predator-catches-and-releases.html.

Marquis, Kim. "Juneau's Famous Black Wolf Mysteriously Disappears." *Juneau Empire,* January 30, 2010. www.adn.com/2010/01/29/1116985/juneaus-famous-black-wolf-mysteriously.html#ixzz1VmdDaYOJ.

Morrison, Eric. "Officials Fear for Romeo." *Juneau Empire,* February 8, 2007. http://juneauempire.com/stories/020807/loc_20070208018.shtml.

MSNBC.com, via Associated Press. "Romeo, Black Wolf of Juneau, Back for the Winter." December 26, 2008. www.msnbc.msn.com/id/28381818/ns/travel-seasonal_travel/t/black-wolf-juneau-back-winter/#.TuZjY1bfWSp.

Sterling, Libby. "Juneau Residents Considered Black Wolf Romeo a Friend." *Fairbanks Daily News-Miner,* June 5, 2010. www.newsminer.com/view/full_story/7804925/article-Juneau-residents-considered-black-wolf—Romeo—a—friend—#ixzz1Vym0TuiK.

Stolpe, Klas. "Juneau Man Receives Suspended Sentence for Hunting Violations." *Juneau Empire,* November 4, 2010. http://juneauempire.com/stories/110410/loc_730859127.shtml.

Weckworth, Byron V., Natalie G. Dawson, Sandra L. Talbot, Melanie J. Flamme, and Joseph A. Cook. "Going Coastal: Shared Evolutionary History Between Coastal

British Columbia and Southeast Alaska Wolves (*Canis lupus*)." *PLoS ONE* 6, no. 5 (2011): e19582. doi:10.1371/journal.pone.0019582.

KIMBERLY GERSON lives for that unique bright-light instant when she is truly and profoundly astonished. She writes about science and nature in order to bring those extraordinary moments and brief encounters to light. Her blog, *Endless Forms Most Beautiful*, can be found at http://kimberlygerson.com.

20

RATS, BEES, AND BRAINS: THE DEATH OF THE "COGNITIVE MAP"

JASON G. GOLDMAN

Humans, just like all other animals, face the same problem every day: How do we get around the world? How do we navigate? If I leave in search of food, how do I find my way back home? If I have a new restaurant to try, I can use my car's built-in GPS navigation to find my way there and back. Most animals, though, can't use Google Maps or GPS. So let's even the playing field: I'm going to leave the house without a pre-specified destination. And when I find a restaurant that looks interesting, I'll stop driving. That's how animals do it when they don't know where they might find their next meal.

Here's one method I might use. First, I'll use the odometer in my car to figure out how many miles I've driven. I'm not going to be driving straight, though, I'll be making turns—so I'll have to use some basic trigonometry to calculate the total distance to my house at each mile.

If I drive three miles, turn ninety degrees, and then drive four miles, I'll know that home is five miles away—even if I don't know what direction I've driven, I can draw a five-mile radius around my position, and I'll know that my house is somewhere along that circle.

That takes care of distance. To figure out the direction I need to drive to get back home, for each mile I drive on my outbound trip, I can figure out the location of the Sun in the sky relative to home's location and my current location, and, again, use some basic trigonometry to figure out what direction I need to drive to get back home. In addition, I'll have to use some sort of internal chronometer—a mental clock—so that I can correct for the changing location of the Sun over time.

Gerbils, geese, chickens, and even desert ants use a similar mechanism. We don't actually sit there with a pad of graph paper and a protractor and recalculate our distance and position in any explicit way—it all takes place below the level of conscious awareness. If an individual knows two of the angles and the distances between three objects, he or she can calculate the third angle and distance, using principles of basic Euclidean coordinate geometry. Some people call this process "dead reckoning"; in the scientific literature, it's referred to as "path integration."

There's a problem, though: path integration is subject to cumulative error. Over the course of one hour, the estimate of distance can be off by as much as 10 percent, and the direction estimate can be off by as much as two degrees. Also, path integration requires continuous memory of one's position relative to home. If an individual becomes disoriented, or forgets, the ability to path integrate becomes useless.

To reorient themselves and regain their bearings, animals need to utilize enduring representations of the environment. For example, animals can rely on their viewpoint-dependent memory of visual scenes. In particular, we can use visual landmarks—items or features of the visual scene that tend not to change over time, such as signs, buildings, or even certain trees. It's one thing to assess whether or not an animal can find its way home. It's another problem entirely to determine the cognitive mechanism in place that allows the animal to get home in the first place. That is the problem that cognitive psychologists must grapple with every day.

Ask most people how they manage to navigate the world, and many will tell you that they think about the space around them in maplike terms. Perhaps they will explain that they imagine a bird's-eye view of their neighborhood. This, they will insist, allows them to get to the store from the office, even if they have never driven that particular route before (they usually drive to the store from home, not from the office).

For most of the past century, cognitive psychologists would have argued that human intuitions happened to be correct when it came to the existence of a "map within the mind." Our story starts in the 1940s, with a group of rats at the University of California, Berkeley, in the laboratory of the experimental psychologist Edward Tolman.

Laboratory rats can quickly learn their way around mazes, and once they do, they can move between individual locations within the maze quickly and efficiently. In the typical maze, a hungry rat is placed at the entrance and wanders around until he or she eventually finds the hidden food, and eats. As this process is repeated day after day, the rat makes fewer and fewer errors and the time it takes to reach the food gets shorter and shorter—presumably because he or she learns the proper routes.

One particularly useful maze that researchers use to test navigational abilities in mice and rats is the Morris water maze. The animal is put into a round tank and it begins to swim. In one spot, there is a submerged platform, which will allow the animal to comfortably stand without needing to swim or tread water to keep afloat.

This would be pretty easy in clear water, so the tank is filled with an opaque milky liquid. The walls of the tank have various geometric markings on them in different places. After being introduced to the maze and finding the location of the platform during training, will the animal be able to use the navigational cues given by the different geometric markers to swim directly toward the platform during testing? The tank is circular, so corners and edges cannot be used to aid in navigation—the only cues available are the geometric markings.

Indeed, rats reliably learn to swim directly for the platform, as long as the geometric markings remain constant. In 1948, Tolman used experiments like this and others to infer that rats and people constructed cognitive maps in their minds.[1] He wrote, "We believe that in the course of learning something like a field map of the environment gets established in the rat's brain . . . And it is this tentative map, indicating routes and paths and environmental relationships, which finally determines what responses, if any, the animal will finally release." The idea of the cognitive map became well entrenched in cognitive psychology. It certainly is intuitive.

Neurophysiological experiments were designed that allowed researchers to monitor the activity of individual neurons in rats while they navigated a maze. Neurons in the hippocampus activate reliably in these sorts of tasks, with each neuron firing in response to a specific place in the environment (these have been called "place cells"). Supporting this theory is the finding that damage to the hippocampus impairs performance on the water maze. By the late 1980s, cognitive neuroscientists came to believe that the cognitive map was stored within the hippocampus.[2]

For more than thirty years, the existence of the cognitive map was generally accepted in psychology. But by the early 1990s, the cognitive map began to fray.

It was certainly possible that the types of mental representations that rats used while navigating were like real maps, as Tolman thought, containing information about the locations of places and the relationship of each place to each other place, independent of the individual's viewpoint. In this sort of model, the cognitive map consists of a "bird's-eye view" of the environment.

There is another possibility, however. It is possible that the mental representations activated by these sorts of navigation tasks are like photographs or snapshots, which simply capture the appearance of each place from a particular point of view. Do rats navigate by using a cognitive map or by using a series of viewpoint-specific visual "snapshots"?

To determine which type of mental representation they used, in 1991 researchers constructed a new kind of water maze. It was already known that, given free access to all vantage points from within the maze, rats could navigate to the platform from anywhere else in the maze. What if the researchers restricted the rats' access to one portion of the Morris water maze during the training phase? This would prevent the rat from forming any viewpoint-specific snapshots from anywhere within the blocked-off space. Then, during testing, they would remove the barrier and drop the rat into the maze in the space where the barrier used to be.

If the cognitive map hypothesis was correct, then the rats should have been able to successfully navigate directly to the platform from the part of the maze in which they had not been able to form any visual "snapshots." However, if rats rely on viewpoint-specific "snapshots" of the visual scene in order to navigate, then when dropped into the new area, they ought to behave as if they're in an entirely new maze. If cognitive maps didn't exist, then the rats should take significantly longer to find the platform. Indeed, rats placed into the unfamiliar part of the maze were far less successful than when they were placed into the familiar parts of the maze.[3]

It seems unlikely that rats would be the only animals to toss aside the cognitive map. Another set of critters that have been studied in depth for navigation are insects—especially ants and bees. When bees are given a task in which they must find some hidden food, do they form a cognitive map or take viewpoint-specific snapshots? In a 1997 study, researchers separated a bunch of bees into two groups, and each had different forms of training in an arena full of objects.[4]

The first group was able to fly around the entire arena until they found the hidden food, as they would in a natural foraging environment. Then, during the test, no matter what side of the arena the bees were released from, they were able to immediately locate the food again. Just like the rats. But a second group of bees was trained to only approach the food from one specific location. When released from the same spot during testing, those bees were successful, as would be expected. What would the bees do who had been trained on one route but were released from a new location? If bees rely on viewpoint-specific snapshots to navigate, then those bees should take much longer to find the food, compared with those tested on the same route that they had trained for. That is exactly what happened.

Both bees and rats learn to navigate a new environment not by forming cognitive maps, but by using viewpoint-dependent snapshots of scenes within the environment. And not just bees and rats; subsequent studies have verified that lots of other nonhuman animal species also have a view-dependent scene recognition mechanism that they use to navigate, rather than a maplike representation.

The next step was to look at humans: do we use cognitive maps, or do we use viewpoint-dependent snapshots?

In 1998, a pair of cognitive psychologists named Dan Simons and Ranxiao Wang came up with a clever experiment.[5] Participants sat down at a table, and on the table was a set of objects: a brush, a mug, goggles, a stapler, and scissors. After they'd viewed the scene, a curtain would lower over the table. Behind the curtain, one of the objects was moved. Then the curtain was raised, and the participants had to identify which of the five items had been moved. The trick was that for some of the trials, the entire array of objects would be rotated as well, though each object would be in the same place relative to each other object (except for the one that had been moved).

If humans use cognitive maps to remember visual scenes, then the rotation of the table should not negatively affect performance: the participants should be as good at noticing which item had been moved when the table was rotated as when the table was left alone. When the table was not rotated, as expected, it was easy for the human participants to indicate which of the five objects had been moved. However, their performance suffered when the table was rotated.

In other words, when looking at the table from a different viewpoint, they were unable to determine which item was in a different spot relative to each other item. Human adults, it appeared, use a view-dependent scene recognition system as well. They take visual snapshots rather than forming a cognitive map. When the viewpoint was changed, the human participants might as well have been looking at an entirely new array of objects. Similar experiments have been conducted with children and even with very young infants. Together, these studies show that this system comes online quite early in life and persists relatively unchanged through development.

If humans indeed have the same cognitive mechanism for representing scenes as bees and rats and other animals, then it is reasonable to ask if this specialized mechanism also has a specialized location in the brain. Studies of rats indicated that representations of scenes are localized in the hippocampus and the surrounding area. If humans form the same sorts of "snapshots" as rats, then similar areas ought to be activated by similar tasks in the brains of both species.

To test this prediction, cognitive neuroscientists Russell Epstein and Nancy Kanwisher conducted an fMRI experiment in which human adults simply viewed pictures of scenes, faces, and objects.[6] What they found was that one particular location in the brain—within the hippocampus—responded robustly to scenes, very weakly in response to objects, and not at all to faces. They named it the *parahippocampal place area*, which has become known as the PPA.

Epstein, Kanwisher, and others did more fMRI experiments to further clarify the role of the PPA. They found that it reliably activated in response to scenes, whether indoor or outdoor, and whether those scenes are empty or full of objects. It did not highly activate, on the other hand, for things that are typically located within scenes (such as furniture within a room) in the absence of surrounding, space-defining surfaces (such as the walls of a room). Taken together, the evidence is overwhelming. The PPA does seem to be the brain region in humans homologous to the "place neurons" in the rat hippocampus.

And so it was that the end of the millennium brought with it the end of the cognitive map (at least in the literal, spatial sense; the term is still used metaphorically for webs of concepts). This is not to say, of course, that humans can't use maps. The critical difference is that our internal representations of the environment are not maplike. Luckily, my GPS doesn't force me to use a snapshot matching system to get around Los Angeles.

NOTES

1. Edward Tolman, "Cognitive Maps in Rats and Men," *Psychological Review* 55, no. 4 (1948): 189–208.
2. John O'Keefe and A. Speakman, "Single Unit Activity in the Rat Hippocampus During a Spatial Memory Task," *Experimental Brain Research* 68, no. 1 (1987): 1–27.
3. Ian Q. Whishaw, "Latent Learning in a Swimming Pool Place Task by Rats: Evidence for the Use of Associative and Not Cognitive Mapping Processes," *Quarterly Journal of Experimental Psychology, Section B: Comparative and Physiological Psychology* 43, no. 1 (1991): 83–103. doi:10.1080/14640749108401260.
4. Thomas S. Collett and Jonathan A. Rees, "View-based Navigation in Hymenoptera: Multiple Strategies of Landmark Guidance in the Approach to a Feeder," *Journal of Comparative Physiology A: Sensory, Neural, and Behavioral Physiology* 181, no. 1 (1997): 47–58.
5. Daniel J. Simons and Ranxiao F. Wang, "Perceiving Real-World Viewpoint Changes," *Psychological Science* 9, no. 4 (1998): 315–20.
6. Russell Epstein and Nancy Kanwisher, "A Cortical Representation of the Local Visual Environment," *Nature* 392, no. 6676 (1998): 598–601.

JASON G. GOLDMAN is a doctoral student in developmental psychology at the University of Southern California in Los Angeles. His research focuses on the evolution and architecture of the mind, and how early experiences might affect innate knowledge systems. In addition to blogging at ScientificAmerican.com, he serves as psychology and neuroscience editor for ScienceSeeker.org, and he edited the 2010 edition of *The Open Laboratory*.

21

DON'T PANIC: SUSTAINABLE SEAFOOD AND THE AMERICAN OUTLAW

MIRIAM C. GOLDSTEIN

Time: 9:00 p.m., after a long day in the lab
Place: Lucha Libre Taco Shop
Internal Monologue:
BAD MIRIAM: "If I do not have a Surf 'n' Turf burrito I will surely perish!"
GOOD MIRIAM: "No! Shrimp is bad! You know shrimp is bad! You are a goddamn marine biologist!"
BAD MIRIAM: "But it is sooooo delicious. Plus it tastes so good with the Super Secret Chipotle Sauce."
GOOD MIRIAM: "Pollution! Bycatch! Habitat destruction! BAD! Bad naughty Miriam!"
BAD MIRIAM: "Shut the hell up while I eat this best of all possible burritos."
NOM NOM NOM.
The End.

As seen in this short glimpse into my psyche, I understand how hard it can be to eat sustainable seafood. I am fully informed as to the environmental cost, and yet I regularly lose my battle against that delicious burrito. It's just really fun to rebel against a smug environmental scold, whether that scold is your annoying vegan cousin or, as in my case, just your own conscience.

The appeal of thumbing one's nose at ever-present environmental guilt is why this "Outlawed Seafood" dinner from Boston restaurant Legal Sea Foods is both brilliant and insidious. Via Grub Street Boston and Garrett Guillotte:

> There might be a panic over seafood sustainability, but Legal Sea Foods CEO Roger Berkowitz isn't taking the bait. Instead, he's hosting a dinner on January 24 with the New England Culinary Guild to address "outdated" scientific findings that turn the dining public against certain species of fish. Behold, a special feast featuring items that people often think are outlawed or blacklisted.

To translate: If you eat this seafood, you are a rebel, an outlaw, going boldly against conventional wisdom. You, like Berkowitz, are not someone who engages in environmental "panic." Leave that for those annoying the-sky-is-falling (or in this case, the-fish-are-disappearing) environmentalists.

So what is on this rebellious outlaw menu? According to Grub Street:

> Fritters
> Black tiger shrimp, duck cracklings, smoked tomato, and avocado sauce
> Cod Cheeks
> Spaghetti squash, toasted pecans, melting marrow gremolata
> Prosciutto Wrapped Hake
> Braised escarole, Rancho Gordo beans, blood orange marmalade

Well, DAMN. That sounds great. Ever since I fell off the kosher wagon I would eat a shoe wrapped in prosciutto, never mind delicious fresh fish. And please sign me up for anything involving duck cracklings.

Except there's one problem. Every single one of these items—black tiger shrimp, Atlantic cod, and Atlantic hake—are listed as AVOID by the Monterey Bay Seafood Watch. Black tiger shrimp, listed under Shrimp (Imported), is classified as AVOID primarily for the incredible habitat destruction that farming it wreaks on mangrove forests and artisanal fisheries, and U.S. Atlantic cod and hake are listed due to severely depleted populations. (Silver and red hake are listed as a "Good Alternatives," but I think Legal Sea Foods is referring to white hake, since it is by far the most popular.)

What does Legal Sea Foods CEO Roger Berkowitz know that

the researchers at Monterey Bay don't? I called Legal Sea Foods to find out.

I spoke to Rich Vellante, executive chef of Legal Sea Foods, who designed the menu around these seafood items. Mr. Vellante said that seafood sustainability is a complex and confusing issue. "In my opinion there's no right or wrong. This is about people trying to educate each other . . . We want to make decisions based on sourcing and not broad-brush everything."

According to Vellante, the cod and hake are locally sourced from the Gloucester and Chatham (Massachusetts) and Portland (Maine) fisheries, and are caught by day boats using hook and line, not trawls. Vellante also emphasized that in the case of the cod, they were featuring the cheek meat, a cut often overlooked by American consumers, so they were using a greater portion of the fish.

In regard to the shrimp, Vellante said that Legal Sea Foods did send people to Vietnam to inspect the shrimp-farming operations and that there were "certain stipulations" that had to be followed, but he was uncertain about the nature of those stipulations and referred me to the Legal Sea Foods marketing department. I called twice but was unable to reach the person he referred me to.

So what does the latest, non-outdated science say about Legal Sea Foods's claims?

Black Tiger Shrimp

According to Seafood Watch, these shrimp should be avoided due to habitat destruction and pollution. Conveniently, a study of the environmental consequences of shrimp aquaculture in Vietnam was published in June 2011. The province discussed in the paper, Can Gio, has undergone mangrove restoration after having been deforested in the 1960s and 1970s, and a section is now a UNESCO International Biosphere Reserve. The growth in area devoted to shrimp ponds has leveled off over the past six years, which means that mangroves are not currently being destroyed for shrimp farms.

However, the study found that significant water pollution results from shrimp farming in Vietnam, and that many farms released wastewater and contaminated sediment that violated Vietnamese water quality standards. Not all farms did this, but on average, shrimp farms' effluent had such high nutrient concentrations that it was similar to agricultural fertilizers. This level of pollution is extremely damaging to surrounding marine environments.

Therefore, while ongoing mangrove destruction may not be a current issue in Vietnam, severe pollution of the remaining and restored mangroves by shrimp farms is an ongoing problem.

Atlantic Cod

When I began researching this, I was surprised to learn that as of the year 2010, Gulf of Maine cod was no longer classified as "overfished" by the National Marine Fisheries Service (NMFS). This is because spawning biomass (basically the number of reproductive-age fish times average weight at maturity) is at half of target levels. Data provided by NMFS looks pretty cheery, since it only looks at cod populations since 1982, but if cod stocks are compared to historical levels, current cod populations are truly pitiful.

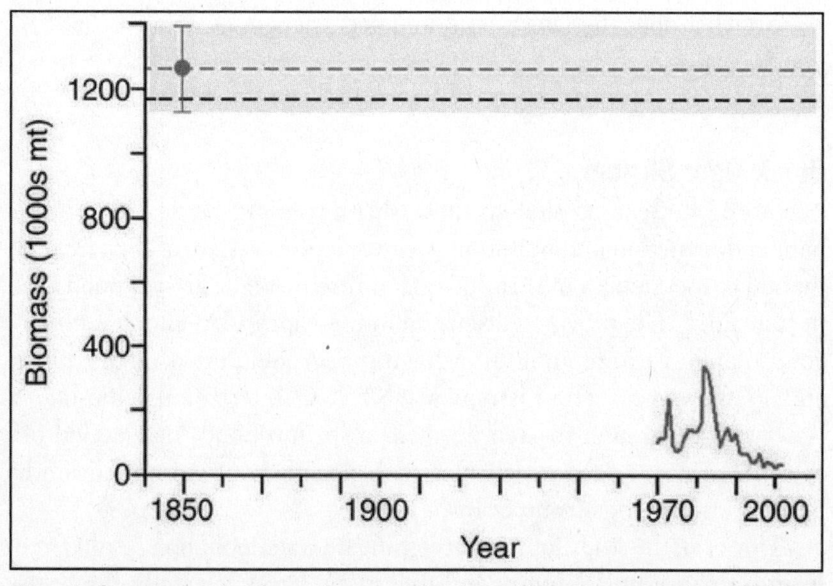

Historical estimate of cod populations. The dotted line is estimated cod biomass on the Scotian shelf in 1852, and the solid line is cod population today. (*Frontiers in Ecology and the Environment* by Ecological Society of America, copyright 2005)

Looking at that, it's pretty hard to argue that cod populations are just fine. Nonetheless, I commend Legal Sea Foods for buying from local hook-and-line fishers. Trawling is undeniably damaging to benthic ecosystems.

Hake

As of 2006, hake was in a pretty sad state. This graph of biomass is from the National Oceanic and Atmospheric Administration (NOAA) stock assessment, showing that hake populations are at less than half of what they were in the 1980s.

As of 2006, NOAA found that hake was overfished, and that overfishing was still occurring. I was unable to find a more recent assessment.

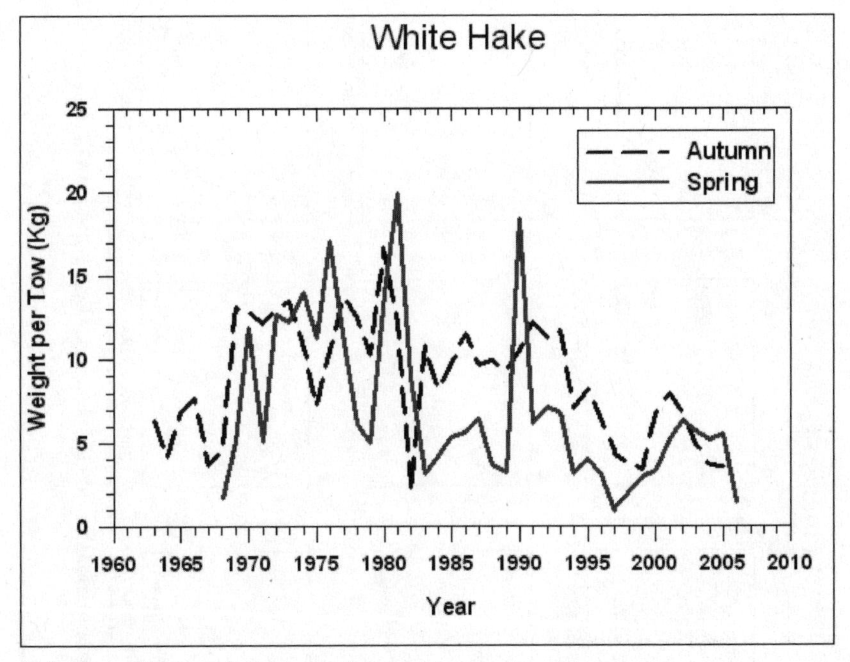

Hake from 2006 stock assessment (Courtesy: NOAA Fisheries)

Now for the Big Picture

The only case of "outdated scientific findings" I could find was that Gulf of Maine cod is no longer classified as "overfished." I personally

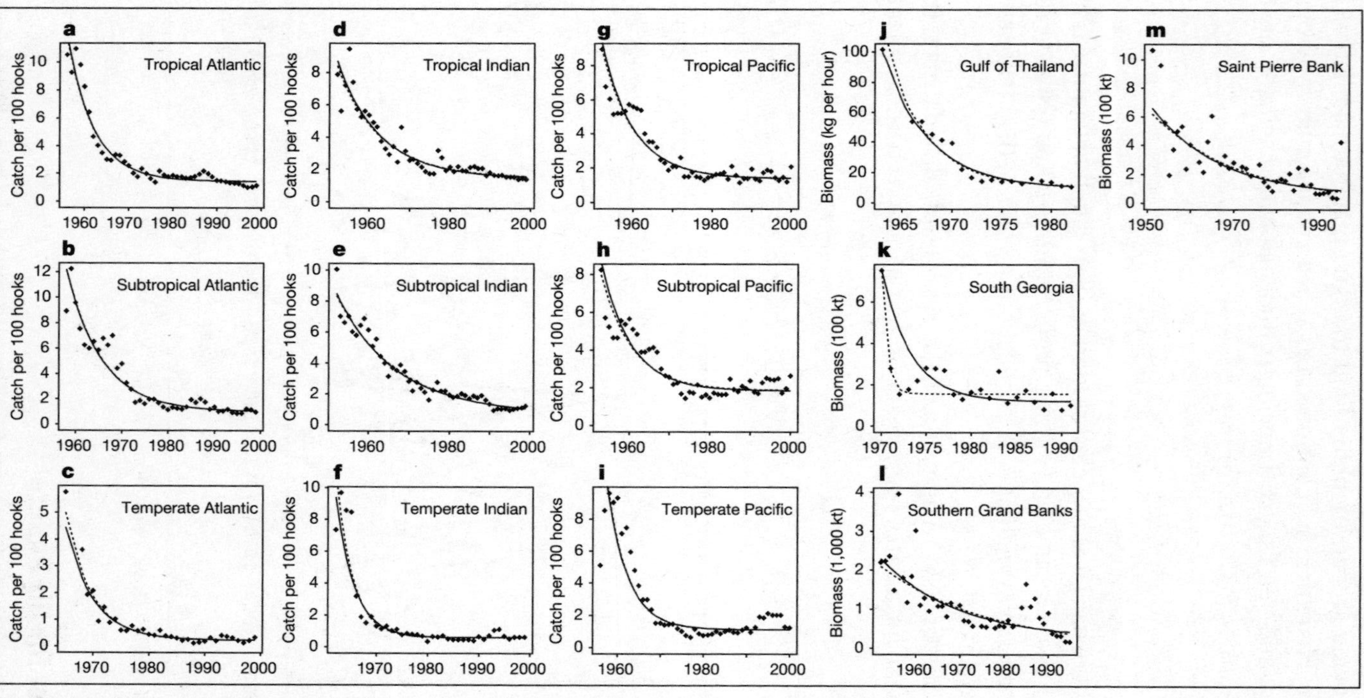

Decline in fish populations worldwide (Ransom A. Meyers and Boris Worm, Nature Publishing Group)

would not feel guilty about eating locally caught Gulf of Maine cod cheeks (yum!), but would continue to attempt to avoid black tiger prawn and hake. (Addendum: As of December 2011, based on new data, NOAA has determined that Gulf of Maine cod are indeed being overfished, and are considering increased restrictions on the fishery. The most current stock assessment had not been publicly released at the time of this writing.)

And here's the big picture, as presented in Myers and Worm (2003). This is biomass data from the beginning of large-scale industrialized fishing in the 1960s. (New England/Eastern Canada fisheries are excluded, as they were already being exploited.)

There are just less fish than there used to be. This means that we need pressure from consumers for effective management. People have to have the correct information about where their fish comes from, and to understand why they should care. The United States has made great strides in effective fisheries management, which should be celebrated, but thumbing one's nose at those no-fun fisheries scientists and environmentalists is not going to change the fact that many fisheries are in serious trouble.

In our conversation, Vellante summarized the purpose of the Legal Sea Foods "Outlawed Fish" dinner by saying, "I think there are a lot of questions and answers to be had. I don't think there is one sweeping answer to everything. We want to create some dialogue around that." Unfortunately, the data show that many fisheries worldwide do have one thing in common: a downward slide.

REFERENCES

Anh, Pham Thi, Carolien Kroeze, Simon R. Bush, and Arthur P. J. Mol. "Water Pollution by Intensive Brackish Shrimp Farming in South-East Vietnam: Causes and Options for Control." *Agricultural Water Management* 97, no. 6 (2010): 872–82. doi:10.1016/j.agwat.2010.01.018.

Myers, Ransom A., and Boris Worm. "Rapid Worldwide Depletion of Predatory Fish Communities." *Nature* 423, no. 6937 (2003): 280–83. doi:10.1038/nature01610.

Rosenberg, Andrew A., W. Jeffrey Bolster, Karen E. Alexander, William B. Leavenworth, Andrew B. Cooper, and Matthew G. McKenzie. "The History of Ocean Resources: Modeling Cod Biomass Using Historical Records." *Frontiers in Ecology and the*

Environment 3, no. 2 (2005): 78–84. doi:10.1890/1540-9295(2005)003[0078:THOORM]2.0.CO;2.

Sosebee, Kathy. "Status of Fishery Resources off the Northeastern US: White Hake." NOAA Northeast Fisheries Science Center. Revised December 2006. www.nefsc.noaa.gov/sos/spsyn/og/hake/.

MIRIAM C. GOLDSTEIN is a Ph.D. student in biological oceanography at the Scripps Institution of Oceanography. She blogs at *Deep Sea News* (http://deepseanews.com) and tweets at @MiriamGoldste.

22

ON BEARDS, BIOLOGY, AND BEING A REAL AMERICAN

JOE HANSON

I've never been able to grow a beard, thanks to my fair-haired Northern European ancestry. Scientifically speaking, it really just means that my dermal papillae don't produce enough 5-alpha-reductase, which isn't my fault. Instead, I produce what could best be replicated by gluing patches of thin, patchy stubble to your face with the lights turned off.

But so many cool dudes in bands sport beards. They were everywhere at this year's SXSW music (and arts and interactive and style) festival in my hometown, Austin, Texas. Beards on fans, beards on drummers, beards on guitarists, beards on DJs . . . beards everywhere. There was even a beatboxer named Beardyman. You show me a band without at least one beard and I will show you a liar.

I have always been a little bit jealous. But then I stumbled upon an Ig Nobel–winning work of microbiology from back in 1967, when men were men and being able to grow a beard was a prerequisite for owning property in thirteen states.*

"Microbiological Laboratory Hazard of Bearded Men"
Manuel S. Barbeito, Charles T. Mathews, and Larry A. Taylor
Applied Microbiology 15, July 1967

Although I am far from the first to critique this research, we must again ask: Does a bearded scientist put his friends, family, and coworkers

* Not actually true.

in danger solely by possessing full and lush chin hair? Is a face-blanket a potential carrier of infectious microorganisms from bench to bedside? A few isolated cases of Q fever had then been confirmed as a result of handling the laundry of exposed laboratory workers, so a resurgence of beards in biology was certainly worthy of scrutiny at the time. The authors made their hypothesis very clear: "After many years of absence from the laboratory scene, beards are now being worn by some persons working with pathogenic microorganisms . . . [I]t has been our policy that beards are undesirable because they may constitute a risk to close associates." I think the true beauty of this research lies in their two-pronged approach to the question.

In the first series of experiments, the authors tested whether certain exposures and beard-washing methods could influence the persistence of bacteria in the hirsute neck-jungles of four volunteers (two of whom were also authors). These men sported oddly age-specific seventy-three-day-old beards and were sprayed with two different types of bacteria: *Serratia marcescens* and *Bacillus subtilis* var. *niger*.

Two intervals between beard-spraying and beard-washing were chosen for the study. Notably, the short interval was designed to simulate "the time necessary for a man to complete a laboratory operation in a zealous attempt to avoid loss of an experimental series despite a known accidental contamination of his beard," because only the truly zealous among us would jeopardize the safety of our friends and facial hair to complete an experiment. The participants were then subjected to a variety of beard-washing techniques.

Post-lavage, various collection methods were employed to analyze the persistence of microorganisms on washed and unwashed beards, which I expect could best be imagined by picturing a room full of Ulysses S. Grant impersonators combing through their facial hair while hovering over a petri dish.

Analysis of the reported bacterial colony counts leaves no doubt that bacteria survive quite diligently on a washed beard when compared to naked skin. It is worth noting, however, that one would effectively need to dip one's beard in a bacterial culture to obtain the exposures used in the study. You have been duly warned.

Bacteria are not the only dangerous microorganisms common to the modern laboratory. In order to study the ability of infectious virus particles to be transmitted by beard, the team employed a set of methods that can only be described as groundbreaking . . . or insane. It turns out that in 1967 you could purchase natural hair beards from Piscataway, New Jersey.

One of these beards was placed upon the face of a mannequin and summarily doused with Newcastle disease virus prepared at high titer. The scientists then stroked live baby chickens across the beard to assess viral transmission.

Unsurprisingly, when a live chicken is rubbed across an unwashed beard containing a lethal titer of avian viral particles, and is then ground up in a blender and injected into fertilized eggs, the rates of survival are not good. Beard-wearing scientists must take care to ensure that they do not repeat this extremely precise and odd sequence of events, lest they ruin dozens of perfectly good eggs.

The authors came to the conclusion that "a bearded man is a more dangerous carrier than a clean-shaven man because the beard is more resistant to cleansing." However, I have been unable to come up with a realistic laboratory scenario that could result in a high titer of deadly virus being sprayed directly on a man's beard without either killing him or destroying the lab, so I am forced to come to a different conclusion than the authors.

In 1967, the United States was undergoing what could be described as "extreme cultural paradigm shifts." This was evident in more-than-minor historical events such as the civil rights movement, free love, and Vietnam War protests. It is not completely unrealistic that a health and safety office stationed at the Fort Detrick army base would hold Bearded Americans of the Counterculture in low esteem. It is common knowledge that the more liberal and progressive elements of American culture at the time were pretty high on beards.

So maybe there was a cultural undercurrent to this study, intentional or not. It would serve the purposes of "The Man" quite well to have some scientific basis for rejecting the beard and its associated underpinnings. Then again, we must consider the off chance these scientists were well aware of the tumultuous stigma of the beard in the United States, circa 1967. One could imagine this study as a well-crafted jab at the white male establishment of government-funded institutional science by a small group of intelligent rabble-rousers. It would certainly be an effective poke, and one that I think still resonates today.

Have the tables been turned today? Scientists are now so often perceived as the ones acting in opposition to American values. There's an interesting double standard there, when you think about it. As I observed at South by Southwest, beards are considered pretty damn cool when it comes to musicians, and people love musicians. Perhaps we can all overlook the inherent infectious-disease risks of science beards as claimed by Barbeito et al. and, by increasing their frequency, increase the acceptance of scientists in today's society.

Whatever the case, I still can't grow one. At least I have a full head of hair to fall back on. Now if you'll excuse me, I have some 5-alpha-reductase to order from a South Asian online pharmacy.

JOE HANSON is a molecular biologist and Ph.D. student from Austin, Texas. He is passionate about bridging the gap between working scientists and the people who are affected by their research. He publishes *It's Okay to Be Smart* (www.itsokaytobesmart.com), a top science Tumblr, and is on Twitter at @jtotheizzoe.

23

I LOVE GIN AND TONICS

MATTHEW HARTINGS

I have a confession to make. I really don't like gin. There is something about that flavor that doesn't sit right with me. I tend to prefer vodka martinis. Blasphemy, I know. If I do have a gin martini, I want more than just a hint of dry vermouth in my drink. I also require olives. Copious olives.

In that same vein, I don't like tonic water either. In fact, I abhor tonic water. It is too bitter for me. I just don't understand how people can drink tonic water. I realize that I'm probably oversensitive (a supertaster, if you will). I want to like each of these drinks; believe me, I do.

My noted aversions to both gin and tonic make it all the more incredible that I love gin and tonics. I can't remember when I had my first gin and tonic. I just remember it being a revelation. A good gin and tonic, to me, is wonderfully crisp, strangely sweet, and aromatic. While it is considered a summer drink, I am perfectly happy to have one in hand any time of year.

Why do I like gin and tonic combined but dislike both gin and tonic separately?

There must be some scientific explanation for this, right? Of course there is. It all comes down to what gin and tonic each look like at the molecular level.

Let's start with the gin. Gin is typically a grain alcohol that has been redistilled with juniper berries or other natural flavorings (citrus peel and other spices). While the alcohol itself lacks much flavor (think vodka), the primary flavors attributed to gin are those from the juniper berries. During the distillation, the alcohol is able to draw several oils—flavors—out of the cells in the berries. The primary flavor oils of the juniper berry look like the following:

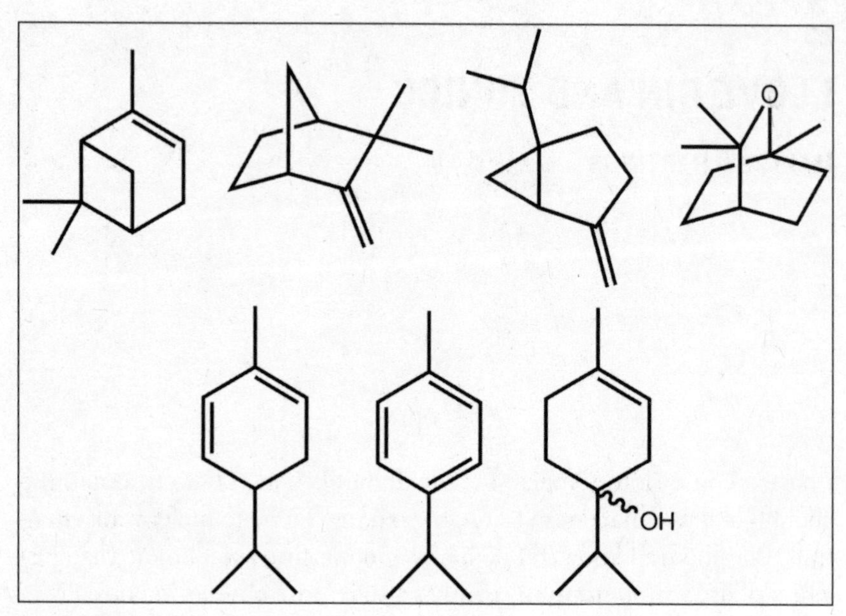

(Matthew Hartings)

At the top, from left to right, we have pinene, camphene, sabinene, cineole. At the bottom, also from left to right, we have terpinene, cymene, terpinen-4-ol. If you're not a chemist, the names aren't important. The structures, however, are important. These are the molecules that give junipers, and gin, their distinctive flavors and aromas.

Tonic water is flavored with quinine. Quinine tastes bitter. It is a base (the opposite of an acid). Quinine was used to treat malaria from the 1600s all the way through the 1940s. It was the British living in India who first mixed quinine tonic with gin to make tonic more palatable. Humans' general aversion to bitter foods likely evolved due to the fact that many poisons are also bitter. Even quinine is poisonous in large enough doses. (Don't worry, your tonic water is perfectly safe.)

On the next page is the chemical structure of quinine. Quinine is a basic compound. Mixtures of water and quinine have a high pH. Humans experience basic liquids as having a bitter taste.

When gin and tonic are mixed, quinine and the flavor molecules from the juniper berries combine to make a perceived flavor that is different from just the sum of the individual parts. The molecules from the

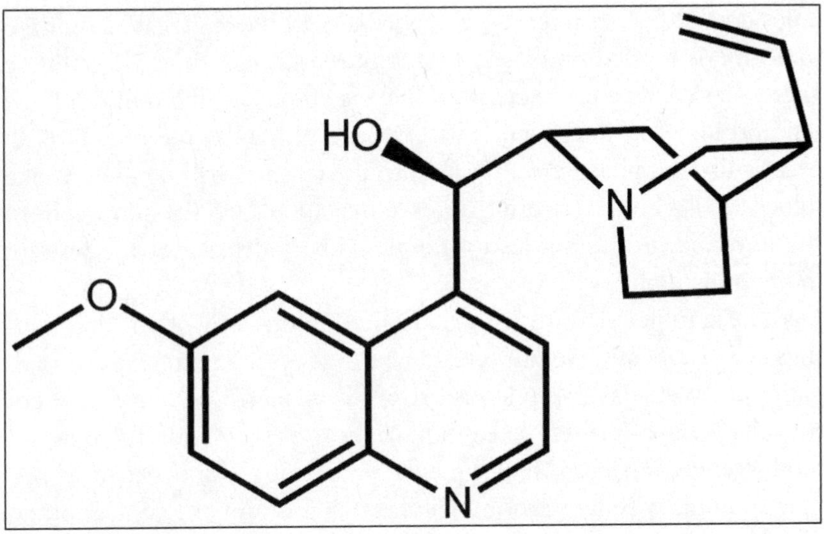

(Matthew Hartings)

gin and the tonic can do this because they look alike; the molecules are similar.

That's right: molecules are homosexual. Molecules that are alike are attracted to one another. Molecules that look nothing alike tend to stay away from one another. You are all familiar with this phenomenon. We have all seen water and oil separating. This happens because water molecules are nothing like fatty oil molecules. After mixing, the oil molecules come together, forming droplets amid all of the water.

Oil and water separate upon mixing because they are molecules that are not similar. Gin and tonic operate on this same principle, only in reverse. The difference is that the flavor molecules from the gin and tonic are attracted to one another. They are similar.

If you look at the gin flavorings and quinine molecules in a new light, you might see that parts of the quinine molecule look like the different flavor molecules from the gin. The juniper oils drawn in gray appear similar to the portion of quinine shown in gray. And the juniper oils drawn in black appear similar to the portion of quinine drawn in black. The parts of these molecules that look alike are attracted to one another. When they are mixed together in a gin and tonic, the molecules come together to create an aggregate. In the aggregate a quinine molecule is nestled up closely to one of the juniper molecules. These aggregates create a taste sensation that is completely different from just gin and tonic on their own. This is why a gin and tonic doesn't taste exactly like gin plus tonic. (This is also the reason why I don't do vodka tonics. Vodka has no flavoring other than alcohol, and the quinine from the tonic has no molecules to mingle with, which leaves the taste of tonic in my mouth.)

These molecular attractions are the reasons why other food pairings are so appealing to our senses. Martin Lersch (a professional inorganic chemist) has compiled a list of flavor pairings—some of them intuitive, some of them seemingly odd—on his wonderful blog on food science, *Khymos*. All of these flavor pairings "work" because each pair contains specific flavor molecules that are similar to one another. The attraction of these molecules alters their flavor profiles. This effect was captured beautifully by a scene in the movie *Ratatouille*, where Remy creates a new flavor sensation by having a bite of strawberry and cheese at the same time.

Thankfully, we still have some bottles of tonic at home and some Bombay Sapphire (one humble man's preference). I plan on going home and putting my knowledge of chemistry to good use this weekend.

Cheers!

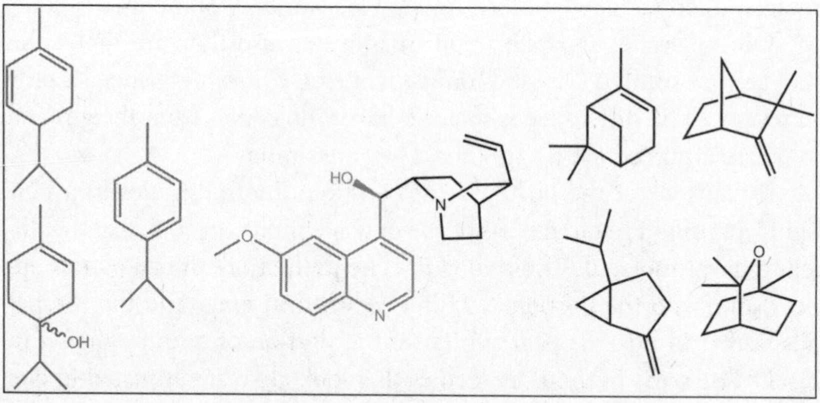

(Matthew Hartings)

MATTHEW HARTINGS is an assistant professor of chemistry at American University, where one of the courses he teaches is titled The Chemistry of Cooking. One of the primary goals of his writing is to engage his readers with the chemistry that they are involved with in their everyday lives.

24

ADORERS OF THE GOOD SCIENCE OF ROCK-BREAKING

DANA HUNTER

"Make them like me adorers of the good science of rock-breaking," Charles Darwin told Charles Lyell once, long ago.[1] This from a man who also once said of Robert Jameson's lectures on geology and zoology, "The sole effect they produced on me was the determination never as long as I lived to read a book on Geology."[2]

That, of course, was before Adam Sedgwick lectured him in geology and took him out for fieldwork,[3] which seems to have done the trick. He did read another book on geology, Lyell's *Principles of Geology*, which became his constant companion on his voyage with the *Beagle*. The concepts of geology prepared him to think in deep time. Without his passion for geology, without deep time sinking deep in his mind, the theory of evolution that changed the world might not be Darwin's.

I have become, like Darwin, an adorer of the good science of rock-breaking.

It's a love that bloomed late. It's always been there, since I was little and wondered at the mountains rising in my back window; at the vast chasm in the ground that revealed billions of years; at the sea that had become fields of stone. But just a bud, tucked away, unopened.

I thought I knew what I wanted and needed from life: a degree in some sort of writerly discipline, like English or maybe history, until I decided the additional debt I'd have to take on wouldn't teach me any more than I could teach myself, and I left academia for the world of daytime wage slavery and nighttime scribbling. I set geology aside, because what a fantasy writer needed couldn't be found in earth and stone.

Dana Hunter, Lockwood DeWitt, rocks, and rock hammer on Doherty Ridge (Courtesy of Cujo359)

So I thought. I searched the stars, delved into physics, waved fondly to geology on my way to geography. I knew the basics: plates moved, mountains rose where they crashed together. Enough to determine the shape of an imaginary world, wasn't that?

No.

And there was the small matter of a subduction zone, now: I'd moved away from Arizona's fossil seas to Washington's young cordillera. I didn't understand this terrible and beautiful new place. It wasn't a landscape I'd grown up with. So I explored it a bit, and the more I explored, the more I needed to understand, the more I realized a story world should be so much more than an ocean with a few haphazard continents sketched in. I wanted to understand this world so that I could understand that.

So I delved, deep, into deep time, into continental crust and ocean floor. I turned to books on geology. They weren't enough. I found a few geobloggers. They were more, still not enough. I began writing about geology in order to understand it, because there's no better way to learn than by teaching someone else. And it still wasn't enough.

The more I learned, the more I realized I didn't know.

And that isn't precisely the problem. If it was, I could decide that knowing a little more than most is quite enough to be going on with, and settle down, content with my little gems of knowledge. If I'd just stayed a bit more ignorant, it would have been okay.

There's a metaphor that explains why those few shining gems, no matter how many more I acquire, will never be enough. It's in the story I'm writing right now, in which Nahash, the Serpent of the Elder Tree, is tasked with giving knowledge and wisdom to a young girl. And this is what he does, the first time they meet:

> He led her round the tree, to the spring that bubbled out from between the roots, clear and deep. Another branch hung low there, and there was fruit on it, so heavy and ripe it was ready to fall. He plucked one of the fruits and turned back to her. "This fruit is knowledge. Do you see? It's probably sweet. Could be sour. You won't know until you've tasted it." He held it out. She reached for it, but he pulled it back. "There's something else. Once you've tasted it, no matter whether it's sour or sweet, you'll always be hungry. You'll starve. And that water, there"—he waved at the spring—"sweetest water in the world, maybe the whole universe, but once you've had a drink from it you'll always be thirsty. Starving and parched. Is that how you want to spend your life? There are other ways of living, you know, and some of them are no less worthy. Some of them are even fun. Or so I've heard."
>
> She held out her hand, but didn't speak.
>
> "Are you quite sure? Because there's no going back, you know. Not ever."

Should I ever become a famous speculative fiction author, people will accuse me of being autobiographical. And, aside from the fact that I was an adult when I ate that fruit and drank that spring water, and didn't actually munch unidentified fruit and drink from the spring of an actual World Tree Serpent, they'll be quite correct. This is completely autobiographical. Since taking a bigger bite and a deeper drink from the fruit and springs of science, especially geology, I've been starving and parched. I'm desperate enough for more that I've considered going deep into debt for a degree I may never earn a living from. I'd beggar myself to get a full meal, and I know I'd walk away with a $30,000+ tab, and I'd still be starving. Add several fistfuls of dollars for grad school, and I'd still feel I hadn't had more than a bite to eat and a drop to drink.

There's no going back, now I'm an adorer of the good science of rock-breaking. There's no end to it, you see. It's a vast old Earth, and there's no way for any of us to know everything about it. And even if we could, have a look out in space—lots more planets out there, all unknown, all fascinating, all with incredible rocks to break.

Geologist Anne Jefferson of the University of North Carolina in Charlotte asked, "If you are a geology enthusiast but not professional . . . what do you wish you could get in additional formal and informal education? What would you like from geosciences students, faculty, and professionals that would make your enthusiasm more informed and more fun?"[4]

And these are the things I'll say to you professionals and pending professionals, you professors and students, you who have careers at surveys and for companies:

Do not withhold your passion.

If there's a book within you, write it. Let your love pour onto the page. Put as much of your knowledge and wisdom into words as you are able, and get it into my hands. You don't even need a publisher in this digital age: you can upload it as an e-book. I'll take whatever you've got. And if you need a wordsmith's help, well, you know where to find me.

If something fascinates you, blog it.[5] Even if it's complicated and you think it's of doubtful interest to anyone outside of the geotribe, post it up there where I can see it. If you love it enough to spend time explaining it, chances are I'll love it enough to spend time doing my best to comprehend it.

If you've written a paper, share it. Blog about it, maybe even offer to send me a PDF if you can. There's a huge, expensive double barrier between laypeople and papers: the language is technical and hard, and the journals charge so much that even if we're willing to put in the work, we may not have the funds. We've already spent our ready cash on books and rock hammers and various, y'see. But if you're allowed to send out a copy, and you can give me an iota of understanding, I'll read it, struggle with it, combine it with those other precious bits of knowledge until I've made some sense of it.

Show me what you see. Post those pictures of outcrops. If we're in the same neighborhood with some time to spare, put those rocks in my hands. I know you've got a career and a family and can't lead many field trips, but if you can take even a few of us out, do it. We'll happily keep you in meals, beer, and gas money just for the chance to see the world through your eyes, in real time and real life.

Answer questions as time allows.

Point us at resources.

Let us eavesdrop on your conversations with other geologists and geology students.

And hell, if you want to make some spare cash, and you're not in a position where there might be a conflict of interest, consider teaching some online classes for a fee. There's plenty of us who can't quite afford college but could scrape together some bucks for the opportunity to learn something directly from the experts. We'd practically kill for that opportunity, but the days when you were allowed to break rocks in prison are pretty much over, so there's not quite as much incentive to break the law.

In other words, mostly do what you're doing now, with maybe a few added extras.

That's what those of us without the cash for a college degree and not even a single community college class on offer need. We just need you to share as much as you can, challenge us as much as you can.

And you there, with the students: make them, like me, adorers of the good science of rock-breaking. Send them out into the world with passion, a hammer, and a desire to babble to the poor starving, parched enthusiasts hoping for just one more bite to eat and drop to drink.

NOTES

1. Charles Darwin to Charles Lyell, August 9, 1828, in Darwin Correspondence Database, www.darwinproject.ac.uk/entry-424; accessed on February 14, 2012.

2. Charles Darwin, "Recollections of the Development of My Mind and Character" [Autobiography]: 19, The Complete Work of Charles Darwin Online, http://darwin-online.org.uk/content/frameset?viewtype=side&itemID=CUL-DAR26.1-121&pageseq=50.
3. AboutDarwin.com, "People who influenced Darwin," last modified September 4, 2011, www.aboutdarwin.com/people/people_01.html#0185.
4. Anne Jefferson, "Call For Posts, Accretionary Wedge #38: Back to School," *Highly Allochthonous*, September 8, 2001. http://all-geo.org/highlyallochthonous/2011/09/call-for-posts-accretionary-wedge-38-back-to-school/.
5. Lockwood DeWitt, *Outside the Interzone* (blog), http://outsidetheinterzone.blogspot.com/.

DANA HUNTER is a Seattle-area blogger and speculative-fiction writer who has, under protest, been called a geologist by professional geologists. She's decided it's wise not to argue this point with scientists wielding rock hammers. She loves to build worlds and break rocks. Find her at http://freethoughtblogs.com/entequilaesverdad and http://blogs.scientificamerican.com/rosetta-stones.

25

"I FEEL THE PERCUSSIVE ROAR ON THE SKIN OF MY FACE"

KAREN JAMES

It begins when the white noise of nervous chatter around me gradually coalesces into a ragged chant. "Ten! Nine! Eight!" Then the chant becomes clearer and louder. "Seven! Six! Five!"

Someone to my left shouts, "Main engine start!" I focus hard on the launchpad three miles away. For a second nothing happens and it's as if I'm looking at a still photograph, or even at the wrong spot. The launchpad could be empty if it weren't for a hint of orange—the top of *Atlantis*'s external fuel tank—peeking above gray scaffolding.

What happens next erases all doubt: a white cloud of smoke and steam appears, growing fast from the bottom right side of the pad. This is the plume from *Atlantis*'s main engines. Then an even bigger cloud jets out to the left, this one from the solid rocket boosters (SRBs). Slowly at first, a colossal weight of metal levitates into view above the scaffolding. When the familiar shape of the space shuttle appears, my thoughts lurch toward the four humans on board, their courage and their vulnerability.

As the "stack" (the shuttle together with the external tank and the SRBs) speeds upward, the flame coming from the main engine and the SRBs is unexpectedly bright—brighter, it seems to me, than the Sun. It's a tremendous explosion, after all, and only barely controlled, as the *Challenger* disaster so tragically brought home.

Accelerating toward the thin cloud layer above, the stack rolls over so that *Atlantis* is obscured, from my vantage point at NASA's press site, by the external tank and SRBs. It is a planned maneuver that positions the shuttle in a head-down attitude, to achieve orbit but also to minimize the stress of acceleration to *Atlantis* and her crew.

Four months before the final flight of space shuttle *Atlantis*, space shuttle *Discovery* lifts off for the last time on February 24, 2011. (Photograph by Paolo Dy)

Only when the stack nears the clouds does the sound arrive. It's a rumble at first, corresponding to main engine start; it's loud but not overwhelming. When the sound of the SRBs arrives, however, I not only hear it, I feel the percussive roar on the skin of my face, my chest, my arms. I am physically connected to *Atlantis* now.

All too fast for me, but right on schedule for *Atlantis*, she pierces the clouds. For a few seconds I can still see the bright spot even through the thin layer, but as that fades, I start to become aware of the column of smoke tracing the shuttle's ascent. It's drifting to the left of the launchpad but, remarkably, maintains its shape right down to individual billows.

As my bubble of attention continues to expand, I notice a dark line lengthening from the point where the column of smoke meets the clouds. I realize this is the shadow of the portion of the column that is above the clouds, and that I can monitor the shuttle's progress by watching the shadow extend farther and farther to the left. I can even see it curving slightly as the shuttle begins to curve toward its orbital track.

I start to notice the other people around me again. I'm reassured to see I'm not the only one catching my breath and wiping away tears. The excited chatter returns, this time mixed with sniffles and sighs. Friends hug and pat each other on the back and I hear snatches of conversation:

"Godspeed *Atlantis*," "I can't believe it's over," and even a heartbreaking "My job is now obsolete."

A cheer goes up to celebrate the announcement of "MECO," or main engine cutoff. *Atlantis* and her crew are now in orbit around Earth, traveling at a speed of about 17,500 miles per hour.

In two days' time, they will rendezvous with the International Space Station and get to work on the objectives of their mission: to stock the station with a final delivery of supplies and equipment. The final space shuttle mission is not over; it has only just begun.

KAREN JAMES is a biologist at Mount Desert Island Biological Laboratory in Maine and director of the HMS Beagle Project, which aims to rebuild HMS *Beagle* as a flagship for science. It was through a connection between this project and both NASA and *The Guardian* that she was able to witness the final launch of the space shuttle program from the press site at Kennedy Space Center. Follow her on Twitter at @kejames.

26

FREEDOM TO RIOT

ERIC MICHAEL JOHNSON

From London to the Middle East, riots have shaken political stability. Are the answers to be found in human nature?

Police cars were overturned and shops looted as the mob descended on the city's central square. Rioters tore the police station's outer door off its hinges and "used it as a battering ram" to break inside. Others smashed their way into the city building, where they assaulted government workers, shattered windows, and destroyed furniture.

The portrait of a powerful leader was pulled from the wall and sent dangling from a balcony as angry voices below cursed him and the other "fascists" believed responsible for their condition. One man, a lathe operator who had gone on strike, ran onto the balcony holding up two plates loaded with cheese and sausage. "Look and see what they eat," he shouted to the crowd below, "yet we cannot get such food!"

The Novocherkassk riot on June 2, 1962, was Soviet Russia's largest public uprising since the Revolution. More than two thousand people took to the streets in response to the Communist Party's decision to increase food prices by 30 percent at the same time that wages were being reduced. Workers walked out on the job, students left their classrooms, and men and women of all ages joined the chorus of protest. Marching peacefully through lines of soldiers backed by armored vehicles that had been hastily assembled, the protesters went to voice their grievances directly to a communist government that claimed to be on the side of the worker.

(Nathaniel Gold)

But when authorities, intentionally or not, fired on unarmed civilians in the crowd, the "noisy, aggressive, and far less reasonable members determined its focus and direction," wrote historian Vladimir Kozlov in his book *Mass Uprisings in the USSR*. Drunken fights and petty theft occurred alongside the anger over poverty and police brutality. From a crowd made up of individuals, each possessing the ability to make a free choice, something more powerful had been unleashed in which normal rules of conduct seemed not to apply.

"For some reason some kind of force filled me," testified one of the rioters during his trial. "Until this day, I do not understand how I got into this. What kind of devil was it that asked me to go and forced me to enter into the police department?"

Collective violence, extending from riots to warfare, presents a challenge to our ordinary understanding of free will. Actions that most individuals would rarely take on their own can be embraced by those same individuals when supported by a larger group. This can occur in societies ranging from the communist regime of Soviet Russia to the capitalist free market of modern-day England. Given this commonality, perhaps the collective violence of a riot can be best understood as a biological event in which evolved cognitive responses encounter a unique environmental threat. And if that is the case, do individuals caught up in such incidents have any choice in the matter?

The Evolution of Mob Mentality

"Imagine you're on a bus," explains Vaughan Bell, clinical and research psychologist at King's College London. "It's full of people and you have to jam into an uncomfortable seat at the back." Very little connects you with any of the other passengers, and it is unlikely you would even give them a second thought.

Suddenly, multiple windows are smashed open, and you discover that the bus is under attack by a group of thugs who are trying to steal people's bags through the broken windows. You very quickly feel a common bond with the other passengers and willingly cooperate with them to help fend off the thieves. Extreme circumstances have pushed you into identifying with the group against a common enemy.

"You didn't lose your identity," says Bell, "you gained a new one in reaction to a threat." Bell points out that in the case of riots, that threat is often excessive force from the police that turns a disgruntled crowd into an angry mob.

This scenario is what psychologists refer to as the elaborated social identity model of crowd behavior. Each individual remains a rational actor, but has been primed by natural selection to identify with the group during a period of crisis. This well-developed in-group/out-group bias is what has allowed our species to be the most cooperative of the primates, but certain conditions have the potential to turn us against our own community. While the psychology of collective behavior may explain why individuals join together once a riot is under way, it doesn't explain why the riot would begin in the first place. As it turns out, our primate cousins offer a unique insight into this question. Nonhuman primates offer a window into the range of behaviors available to our evolutionary ancestors and the legacy that they have passed down to us.

"Collective violence," wrote Harvard primatologist Richard Wrangham in the *Annals of the New York Academy of Sciences*, shows "a common human pattern evident in societies lacking effective central authority, manifested in ethnic riots, blood feuds, lethal raiding, and warfare." Such aggression, he says, is directly related to that of nonhuman primates and demonstrates a common evolutionary history. As Wrangham earlier wrote in his book *Demonic Males: Apes and the Origins of Human Violence*, our primate origins "preceded and paved the way for human war, making modern humans the dazed survivors of a continuous, five-million-year habit of lethal aggression."

Wrangham is but the latest in a long history of evolutionary scientists to argue that collective violence is an adaptive feature of the human species. However, one of the earliest case studies to reach this same conclusion is actually, in hindsight, a prime argument *against* this contention. Reconsidering it in this light offers a unique perspective on the factors that motivate collective violence in human societies, and may even offer some clues about how to prevent it.

Anatomy of a Massacre

On October 27, 1930, a brief report in *Time* magazine described an attack that took place among a captive colony of 100 "Abyssinian" baboons that had been installed at Monkey Hill in the London Zoo. This was merely the latest outbreak in what would eventually be described as a massacre of more than two-thirds of the population.

According to the article, the attacks occurred when a young social outcast had "stolen a female belonging to the king." He fled with his captive behind a hastily built barricade, where an indignant crowd gathered, trapping the two inside. After "two days of siege" he attempted to escape from his sanctuary, only to be brutally attacked and killed by the waiting mob. Each attack led to a counterattack by a rival alliance. After several years, the death toll amounted to ninety-four individuals—two-thirds of them male. Among the deceased were fourteen infants, most of them strangled either by male attackers or by their own mothers.

What transpired on that barren landscape was carefully documented at the time by Solly Zuckerman (later Lord Zuckerman), a recent émigré who had just completed his doctorate in anatomy at the University of Cape Town in South Africa. Faced with such senseless brutality inflicted by group members against one of their own, Zuckerman speculated on what could have stirred the hostilities and then kept them going for so long. He eventually concluded that the cause was "social discord." The death of one individual upset the political hierarchy, and violence ensued until stability was regained.

The massacre of Monkey Hill has since become a classic example of a zoological case study that reveals the danger of embracing a faulty assumption about "natural" behavior. Zuckerman assumed he was observing evolution in action, and that natural laws had shaped these beasts to engage in lethal aggression. Like most biologists of the time, he accepted the view that primate social behavior followed a one-size-fits-all model that was unaffected by environmental factors. While human beings learn to rise above their bestial nature, animals are, well, simply animals. Or so the argument went.

"Behavior is uniform," wrote Zuckerman in *The Social Life of Monkeys and Apes*, published in 1932. "The common belief that the new environment [of captivity] grossly distorts the expression of these relationships has no foundation in fact. The pattern of socio-sexual adjustments in captive colonies is identical with that observed among wild animals."

He couldn't have been more wrong. A few decades later ethologist Hans Kummer confirmed this by comparing the behavior of captive baboons in Zurich with wild populations in Ethiopia. What Kummer found was that captive baboons showed many more aggressive acts than their free-ranging counterparts (nine times the average for wild females and seventeen and a half times that for males). The massacre of Monkey Hill therefore represents a kind of controlled experiment on the potential dangers of social engineering, one that demonstrates that flawed assumptions can have lethal consequences.

It is true that Abyssinian baboons, now more commonly referred to as Hamadryas after their scientific designation *Papio hamadryas*, are notorious for their aggressive behavior. But the levels of aggression that Zuckerman described have never been observed in the wild, then or since. Zuckerman assumed that captive conditions had no effect on baboon society, and so didn't think it was relevant that more than 100 individuals were enclosed in a facility that measured only 100 feet long by 65 feet wide. It never occurred to him that the social discord he observed had been manufactured, and that he was the cause. The violence that left nearly 100 baboons dead was the ultimate, and tragic, result.

From Colony to Metropolis

Since the events of Monkey Hill, hundreds of studies with captive primates have shown that impoverished environments result in heightened aggression and antisocial behavior. Such behavior has been shown to increase significantly under conditions of overcrowding; when there's a lack of novelty in food, entertainment, or social opportunities; when the population increases and the number of strangers in a colony grows; or, most crucially, when food is limited and/or fluctuates dramatically (see Honess and Marin 2006 for a review of the literature). Any of these factors can greatly increase the level of stress that individuals experience and can promote social discord.

As neuroscientist Robert Sapolsky has documented in multiple studies, social stress and aggressive responses are highly correlated. When these stressors are too great, what would otherwise be an adaptive response can become exaggerated. Often all it takes is the right conditions to trigger a violent response. But if multiple factors are present and persist, the result can be sustained aggression or even colony collapse. In this way, stress-induced aggressive behavior is both adaptive and the result of environmental conditions—a response that can be exaggerated or distorted when living in captivity.

It should come as no surprise that one of the most important environmental factors involved in stress-induced aggression is food. Another classic, if somewhat cruel, study by Charles Southwick in 1967 found that increasing the amount of food in a captive colony of rhesus macaques by 25 percent decreased the amount of aggression by 50 percent. However, when access to a normal amount of food was restricted (by placing it in a single basket where it could be monopolized by a few high-ranking individuals), the level of overall aggression tripled and the number of violent attacks per hour increased fivefold.

"The animals are especially quarrelsome and competitive when the food supply is restricted," Southwick concluded drily. But he added that "the increased tension and aggressivity of captive animals exaggerates certain types of phenomena, and hence the results must be interpreted in proper perspective with reference to natural situations."

What was clear was that when all colony members had enough to eat, aggression was cut in half. But introducing inequalities, so that only a few individuals enjoyed an excess while the majority went without, met with a violent response. These findings were subsequently confirmed, and now form the basis for food policy in captive primate facilities around the world. Captive conditions are now understood to significantly alter animal behavior, unless precautions are taken both to understand species-typical behavior under natural conditions and to provide an environment that allows its expression. Can the same be said of human societies?

In our own species, historical and sociological studies of factors contributing to collective violence have found some striking parallels with studies of captive nonhuman primates. For example, John Archer in his book *Social Unrest and Popular Protest in England, 1780–1840* showed that major outbreaks of rioting between 1780 and 1822 correlated with high wheat prices. In nearly all cases the riots were preceded by a sharp rise in price, and once the price fell, the incidence of riots fell with it. This isn't to suggest that wheat price alone was the cause, or that a rise in price always resulted in a riot. But it does suggest that the two were correlated, and that a rise in food price promoted the same kind of social discord that lay behind the incidents of collective violence on Monkey Hill.

Identical findings were reported in a study by Marco Lagi, Karla Z. Bertrand, and Yaneer Bar-Yam of the New England Complex Systems Institute in Cambridge, Massachusetts. In their August 11, 2011, report, the authors detail the close correlation between global food prices and the incidence of riots in North Africa and the Middle East. In 2008 more than sixty riots occurred worldwide in thirty different countries during a peak in food prices. After prices declined temporarily in 2009, even higher prices at the end of 2010 and the beginning of 2011 coincided with additional food riots, as well as with the larger protests and revolts that have become popularly known as the Arab Spring. In contrast, there were relatively few incidents of collective violence when food prices were low.

"The timing of peaks in global food prices and social unrest implies that the 2011 unrest was precipitated by a food crisis that is threatening the security of vulnerable populations," conclude the authors. "Deterioration in food security led to conditions in which random events trigger widespread violence."

These findings may also help to explain the timing of riots in England the week before this paper was published. As Bar-Yam and colleagues point out, while the uprisings in their study were associated with dictatorial regimes, rising food prices are likely to affect impoverished communities in otherwise wealthy nations as well.

"The price of bread has more than doubled in the past five years in England," Bar-Yam explained via e-mail. "As in other parts of the world, London has a population of poor individuals vulnerable to [increases in] food prices who are likely to engage in protests and participate in social disorder under these conditions."

Ironically, it may have been the very policies promoted by England and other Western governments that lay behind these conditions. One of the main causes of the spikes in global food price, according to Bar-Yam and colleagues, was investor speculation that resulted in an economic bubble like the one that hit the housing market in 2008. Beginning in 2001, financial institutions like Goldman Sachs and Morgan Stanley in the United States as well as Barclays Capital in the U.K. successfully lobbied their respective governments to deregulate the commodities market. This allowed them to invent new financial products, known as derivatives, which caused an explosion of speculation and volatility in agricultural prices.

According to data from the United Nations, investment in commodity funds rose from $13 billion in 2003 to $317 billion by 2008. The price of food rose along with the value of these investments, creating a financial bubble that put increasing strain on those communities already on the edge.

England's Conservative government implemented austerity measures (at the same time as food prices peaked), which rolled back many social programs poor communities relied on. On November 10, 2010, for example, student protesters rioted when cuts to education caused tuition fees to nearly triple. Likewise, additional cuts targeted youth and community centers, medical coverage, unemployment and disability payments, and child benefits, as well as housing and fuel subsidies for pensioners. England already had one of the most unequal societies in Europe based on the divide between rich and poor. Such austerity measures may have pushed this division to its breaking point.

An interactive map created by the Centre for Advanced Spatial Analysis at University College London makes clear that the riot outbreaks were clustered in the most economically deprived regions of the city. It was these regions that would have been most aversely affected by the austerity measures, and with a peak in both food and energy prices occurring at the same time, the environmental conditions were ideal for a triggering event that would push an already stressed population over into social discord.

This conclusion is further supported by an analysis of similar austerity measures throughout Europe during the twentieth century conducted by economists Jacopo Ponticelli and Hans-Joachim Voth of the Centre for Economic Policy Research in London. According to their report "Austerity and Anarchy: Budget Cuts and Social Unrest in Europe," published in August 2011, there was a "clear link between the magnitude of expenditure cutbacks and increases in social unrest."

As Ponticelli and Voth point out, when expenditure cuts reached 1 percent or more of a nation's gross domestic product, the total number of demonstrations, riots, assassinations, and general strikes in a single year would increase by one-third compared to periods of budget expansion. When budget cuts reached 5 percent of GDP, the number of incidents doubled. According to London's *Financial Times*, England's current budget cuts are 4.5 percent of GDP.

It was in the midst of these environmental conditions that police fatally shot Mark Duggan on August 4, then allegedly beat a sixteen-year-old girl during what was reportedly an otherwise peaceful protest in response to the shooting.

"This was the triggering event," says Bar-Yam, "that led to spontaneous wider violence." After five days of rage, the damage was estimated to be £200 million (more than $300 million) and resulted in the arrest of more than 3,000 people. "It makes sense that in the beginning, the people involved were people in need. The violence then cascaded to others, who took advantage of the social disorder for other reasons. Social disorder is contagious."

As in London, the Novocherkassk riot forty years ago died down as those involved eventually dispersed, sobered up, or found themselves in jail. As the riot population declined, the shared social identity declined with it. But the rioters left behind a physical scar on the urban landscape, evidence of the rage shared by thousands of people during a time of acute environmental stress.

However, while the collective violence may have waned, the political meaning of the events remained hotly contested. Soviet Premier Nikita Khrushchev announced that the Soviet rioters in 1962 were nothing more than "antisocial elements who spoil our lives" and condemned them all as "grabbers, loafers, and criminals." British Prime Minister David Cameron would offer nearly identical words. Others, such as *The Daily Telegraph*'s former editor in chief Sir Max Hastings, would portray the rioters as little different than Zuckerman's baboons from Monkey Hill, "wild beasts" who "respond only to the instinctive animal impulses."

But what is to blame for such cases of collective violence—nature, or the unnatural conditions of modern life? While there may well be evolved responses that promote collective violence, research in captive primates suggest that these behaviors are heavily influenced by environmental stress. During the past year environmental conditions were just right for the triggering of social discord in our own society and, in the contagion that followed, violence quickly spread among a population predisposed to a shared identity.

For London and the cities throughout North Africa and the Middle East, it appears there was a free choice to riot after all. But the choice didn't come from the rioters alone: it rose from leaders and policymakers and the larger society as a whole. Riots reveal a colony in discord. Many of us have acknowledged the widening inequality and the economic decline of our most impoverished citizens—but we chose to ignore it.

Aleksandr Solzhenitsyn, after experiencing firsthand the inequality and injustice that emerged from the Soviet command economy, wrote that the Novocherkassk riot was the first indication that the Iron Curtain was beginning to unravel.

"We can say without exaggeration," he wrote in *The Gulag Archi-*

pelago, "that this was a turning point in the modern history of Russia." Free markets are theoretically designed to be flexible so that they rapidly respond to the needs of individuals and society. If this assumption is flawed, we will need an alternative. Human nature is not destined for social discord so long as we have the freedom to choose conditions that can reduce the potential for collective violence. But the question remains whether we will do so.

REFERENCES

Archer, John E. 2000. *Social Unrest and Popular Protest in England, 1780–1840*. Cambridge: Cambridge University Press.

Honess, Paul E., and Carolina M. Marin. 2006. "Enrichment and Aggression in Primates." *Neuroscience & Biobehavioral Reviews* 30 (3). doi:10.1016/j.neubiorev.2005.05.002.

Kozlov, Vladimir A. 2002. *Mass Uprisings in the USSR: Protest and Rebellion in the Post-Stalin Years*. Translated by Elaine McClarnand MacKinnon. Armonk, N.Y.: M. E. Sharpe.

Lagi, Marco, Karla Z. Bertrand, and Yaneer Bar-Yam. 2011. "The Food Crises and Political Instability in North Africa and the Middle East." New England Complex Systems Institute. arXiv: 1108.2455v1.

Ponticelli, Jacopo, and Hans-Joachim Voth. 2011. "Austerity and Anarchy: Budget Cuts and Social Unrest in Europe, 1919–2009." Centre for Economic Policy Research Discussion Paper. VoxEU: www.voxeu.org/index.php?q=node/6848.

Wrangham, Richard, and Dale Peterson. 1996. *Demonic Males: Apes and the Origins of Human Violence*. Boston: Houghton Mifflin.

Wrangham, Richard, and Michael Wilson. 2004. "Collective Violence: Comparisons Between Youths and Chimpanzees." *Annals of the New York Academy of Sciences* 1036: 233–56. doi:10.1196/annals.1330.015.

Zuckerman, Solly. 1932. *The Social Life of Monkeys and Apes*. New York: Harcourt, Brace & Company.

ERIC MICHAEL JOHNSON has a master's degree in evolutionary anthropology, focusing on great ape behavioral ecology. He is currently a doctoral student in the history of science at the University of British Columbia, where he is looking at the interplay between evolutionary biology and politics. He blogs at *The Primate Diaries*: http://blogs.scientificamerican.com/primate-diaries.

27

INSIDE THE "BLACK BOX" OF NUCLEAR POWER PLANTS

MAGGIE KOERTH-BAKER

On March 11, 2011, a magnitude 9.0 earthquake rocked the eastern coast of Japan and triggered a series of tsunami waves. Within hours, it became clear that the two natural disasters had damaged the Fukushima I and II nuclear power plants, located about four hours north of Tokyo. In Japan, and around the world, people held their breath and waited to see what would happen next. But it's hard to understand how nuclear power plants fail if you don't already know how they work. I wrote and published this story on March 12, with the hope that it would add some context to the increasingly grim headlines. At the time, nobody knew much about what was going on inside the nuclear reactors, and this story reflects that lack of information. But the main point—how the systems work—is still very applicable.

This morning, I got an e-mail from a BoingBoing reader, who is one of the many people worried about the damaged nuclear reactors in Fukushima, Japan. In one sentence, he managed to get right to heart of a big problem lurking behind the headlines today: "The extent of my knowledge on nuclear power plants is pretty much limited to what I've seen on *The Simpsons*."

For the vast majority of people, nuclear power is a black box technology. Radioactive stuff goes in. Electricity (and nuclear waste) comes out. Somewhere in there, we're aware that explosions and meltdowns can happen. Ninety-nine percent of the time, that set of information is enough to get by on. But then an emergency like this happens, and suddenly, keeping up-to-date on the news feels like you've walked in on the middle of a movie. Nobody pauses to catch you up on all the stuff you missed.

As I write this, it's still not clear how bad, or how big, the problems at the Fukushima Daiichi power plant will be. I don't know enough to speculate on that. I'm not sure anyone does. But I can give you a clearer picture of what's inside the black box. That way, whatever happens at Fukushima, you'll understand why it's happening, and what it means.

At a basic level, nuclear energy isn't all that different from fossil fuel energy. The process of generating electricity at a nuclear power plant is really all about making heat, just as it is at a coal-fired plant. Heat turns water to steam, steam moves turbines in the electric generator. The only difference is where the heat comes from—to get it, you can light coal on fire, or you can create a controlled nuclear fission reaction.

A fission reaction is a lot like a table filled with Jenga games, each stack of blocks standing close to another stack. Pull out the right block, and one Jenga stack will fall. As it does, it collapses into the surrounding stacks. As those stacks tumble, they crash into others. Nuclear fission works the same way. One unstable atom breaks apart, throwing off pieces of itself, which crash into nearby atoms and cause those to break apart, too.

Every time one of those atoms breaks apart, it releases a little heat. Multiply by millions of atoms, and you have enough heat to turn water into steam.

In a boiling-water reactor, like the ones at Fukushima, water is pumped through the core—the central point where the actual fission reactions happen. Along the way, fission-produced heat boils the water, and the steam rises up and is captured to do the work of turning turbines.

In the Core

The core is the part that really matters when discussing Fukushima. In the core of a nuclear reactor, you'll find fuel rods—tubes filled with elements whose atoms are unstable, prone to breaking apart and starting the Jenga-style chain reaction.

Usually, the element used is uranium-235. It's refined and processed into little black pellets, about the size of your thumbnail, which are poured by the thousands into long metal tubes. Bunches of tubes—each taller than a basketball player—are grouped together into square frames. These tall, skinny columns are the fuel assemblies.

The fission reactions that happen are all about proximity. In a fuel rod, lots of uranium atoms can crash into one another as they break apart. Pack the fuel rod into an assembly, and lots more atoms can affect one another—which means the reactions can release more energy. Put several fuel assemblies into the core of a nuclear reactor, and the amount of energy released gets even higher.

Proximity is also what makes the difference between a nuclear bomb and the controlled fission reaction in a power plant. In the bomb, the reactions happen—and the energy is released—very quickly. In the power plant, that process is slowed down by control rods. These work like putting a piece of cardboard between two Jenga towers. The first tower falls, but it hits a barrier instead of the next tower. Of all the atoms that could be split, only a few are allowed to actually do it. And, instead of an explosion, you end up with a manageable amount of heat energy, which can be used to boil water.

In Case of Emergency

Now that you understand what's going on inside a nuclear reactor, you get a good idea of what happened at Fukushima. Like the other nuclear power plants in Japan, Fukushima Daiichi got a message from the country's earthquake warning system and shut down in advance of the quake. Basically, that means that control rods—"big metal gizmos," as Charles Forsberg, executive director of the MIT Nuclear Fuel Cycle Project, described them to me*—were inserted into the fuel assemblies, cutting the fuel rods off from one another. But, because you aren't completely separating all the uranium atoms from one another, shutting down the core isn't the same thing as flipping an "off" switch.

When a reactor core is shut down, its energy output drops not to zero, but to about 6 percent of its normal output, Forsberg told me. The reactions grind to a halt over the next few days, as the falling Jenga towers run out of other towers they can actually hit. In the meantime, atoms

* From a telephone interview with Charles Forsberg, March 12, 2011.

keep breaking apart, releasing both heat and fast-moving particles that can penetrate human skin and damage our cells. Because of this, every nuclear reactor has ways of getting rid of the heat and blocking those fast-moving radioactive particles.

When the reactor at Fukushima shut down, it should have been kept cool by water pumped through the core. But, because the tsunami damaged the diesel-powered generators that pumped the water, the core kept heating up. If that sounds like a design flaw, you're right. The Fukushima reactors were built in the early 1970s. In modern nuclear reactor designs, pumps aren't necessary to move water through the core in an emergency shutdown. Instead, the water moves via gravity.

But, in this case, no pumps meant no water movement. So the core got hotter, which boiled off some of the water. The boiling caused pressure in the core to increase. To protect the core, and to prevent a bigger problem, authorities had to vent some of that steam into the atmosphere, which meant venting some of the radioactive particles along with it.

This is also probably tied into the explosion that happened, according to Forsberg.

"There's zirconium in the fuel rods. When you overheat the reactor core, the first thing that happens is that the zirconium begins to react with steam or water and forms zirconium oxide and hydrogen," he says. "You get a mixture of steam and hydrogen. When you release steam into a secondary building [to decrease pressure in the core], the steam condenses and leaves behind just the hydrogen. Then all you need is an ignition source and you can get a hydrogen burn. That's what happened at Three Mile Island. I don't know if that's what happened in Japan, but it's likely to be the source of that explosion."

The good news is that the explosion seems to have happened outside the core. In that case, it's completely reasonable that an explosion could happen without releasing lots of radioactive material. That's because nuclear power plants come in layers, like an onion.

The core is contained within a building that has solid concrete walls, three to six feet thick. It's meant to withstand collision with an airplane. It's also meant to withstand an explosion from inside. But that bunker sits inside something a lot flimsier—a building more akin to a metal shed. It's the shed that exploded today at Fukushima. Because radiation levels didn't rise after the explosion, we can be pretty certain that the bunker is still intact.

How to Win This

This is a serious emergency, but there are some good reasons to be hopeful. According to World Nuclear News and Reuters, there were seven reactors in Fukushima that were affected by the earthquake. Of those, four have access to outside power to run their water pumps and are stable. Three lost their power. Out of those three, one has steady levels of water. Only two have decreasing water levels. But, in recent hours, workers have started pumping in seawater to one of those. Hopefully, both can be stabilized. But it's hard to say right now.

And then what happens? Remember, this is really just an emergency shutdown gone awry. The control rods are still in place. The Jenga columns are still separated. So, over time, the fission reactions will still slow down and stop. As they do, heat levels will drop, and so will levels of radiation.

Really, what we have here is a waiting game. The goal is to keep the reactors stabilized long enough so that the shutdown can completely shut down.

MAGGIE KOERTH-BAKER is the science editor at BoingBoing.net, one of the most-read blogs in the United States. Besides *BoingBoing*, her work has appeared in magazines such as *Popular Science* and *Discover*, and online at websites including National Geographic Daily News and ScientificAmerican.com. *Before the Lights Go Out*, her book about the future of energy in the United States, was published in April 2012.

28

THIS AIN'T YO MOMMA'S MUKTUK

REBECCA KRESTON

Does anyone else have an inordinate fear of canning jams or pickling veggies? Every time I read an article espousing the brine-laden wonders of canning your own homegrown vegetables, I think, "How hard could this really be? I can do this!" And then I hear the niggling voice in the back of my head that whispers, "But what if you get botulism?" And then I mutter in response, "Maybe I'll just buy my own autoclave."

Thinking about the merits of my own pickled spicy asparagus versus an emergency hospital visit reminded me of a story I heard several years ago in a Bacterial Pathogenesis class. In our lecture on *Clostridium botulinum*, the organism responsible for the production of the botulism toxin, our professor noted that several cases of botulism in Alaska Natives occurred as a result of changing methods of fermenting meat. Professor, you had me at "fermenting meat."

Investigating the veracity of this anecdote, I found that the subsistence lifestyle of Alaska Natives (including the Athabascans, Aleuts, and Yup'ik Eskimos) relies extensively upon hunting animals and fishing, and fermented foods have played a large role in their diet and culture for centuries. Such cuisine may include dried or fermented meat of whales and seals, raw muktuk (the skin and blubber layer of whales), and seal oil. Seal oil has been described as "the Eskimo's salt and pepper," giving you some idea of the importance of these animals in their diet. Fish and fish eggs, typically salmon, may also be used to prepare traditional "stink" foods such as fermented salmon "stink heads" and "stink eggs" to be used as condiments with a cheesy-like texture.

The tried and true Alaska Native methods of burying meat underground to ferment had been modified by the introduction of Western conveniences. Tupperware containers and sealable plastic bags were now being used to create a meaty, anaerobic environment that *C. botulinum* was happy to vacation in. Oh plastics, you synthetic polymers, what have you wrought!

I also discovered the staggering statistic that Alaska ranks among the highest in incidence of foodborne botulism in the world. Indeed, nearly half of all cases of foodborne botulism in the United States occur in that icy northern state; the incidence of botulism in Alaska is 8.46 cases per 100,000 compared to Washington's paltry 0.43 per 100,000. Botulism has been repeatedly referred to as an endemic "hazard of the North," but it typically occurs in western Eskimo coastal villages and Native American regions in the southwest of Alaska due to their proximity to aquatic foodstuffs.

C. botulinum is an obligate anaerobe thriving in oxygen-deprived environments, such as the soil. It produces endospores, tough dormant and nonreproductive cells that can withstand nasty environmental conditions. *C. botulinum* is part of the *Clostridium* genus, a deathly clan that includes the charming organisms responsible for tetanus, gas gangrene, and colitis, or inflammation of the colon. Of the latter, the organism *Clostridium difficile* is responsible for dozens of quite nasty, fatal outbreaks within hospitals and nursing homes throughout the past decade. Over the past few years, it has increasingly become more antibiotic-resistant and lethal, and is now the most common hospital-acquired infection in the United States.

C. botulinum is infamously known for producing neurotoxins. There are seven immunologically distinct types of botulism toxins, and toxin type E, strongly associated with consumption of aquatic animals, is most commonly seen in Alaska. All toxins act at cholinergic neuromuscular junctions* by blocking the release of acetylcholine, causing both muscle and nerve paralysis. Nasty stuff.

There is a short eight- to twenty-hour incubation period following consumption of toxin-laden food before the onset of symptoms. Clinical symptoms of foodborne botulism include gastrointestinal, neurological, and muscular signs such as nausea, vomiting, diarrhea, dry mouth, blurry

*Neuromuscular junctions that are cholinergic use the neurotransmitter acetylcholine to transmit impulses, and thus information, between the nerves and skeletal muscle fibers.

vision, difficulty swallowing, and paralysis. Infection initially starts with the toxin blocking the cranial nerve, leading to a descending paralysis that can ultimately lead to respiratory failure and death. Therapy consists of early administration of antitoxin and palliative care to minimize the pain and the worst of the symptoms.

So, back to this intriguing idea of fermenting one's meat. The term "fermented" might be putting it kindly—many ethnographers have described these prepared foods as intentionally putrefied. And, in fact, the fermentation process cannot occur without a carbohydrate substance, so these meats aren't technically "fermented." The ethnographer E. W. Nelson reported the preparation process quite evocatively in 1900:

> Meat is frequently kept for a considerable length of time and sometimes until it becomes semiputrid . . . This meat was kept in small underground pits, which the frozen subsoil rendered cold, but not cold enough to prevent a bluish fungus growth which completely covered the carcasses of the animals and the walls of the storerooms.

The customary preparation process has since been modified from fermenting food in a buried clay pit, enclosed in a woven basket or sewn sealskin (known as a "poke") for weeks or months at a time. Food is now stored in airtight, Western consumer goods such as plastic or glass jars, sealable plastic bags, or even plastic buckets, and eaten shortly after, in a week or month. Additionally, the food many be stored indoors, above ground or in the sun at milder, less optimal temperatures. This move toward storing meat in warmer, anaerobic settings for shorter lengths of time may expedite the "fermentation" process and, subsequently, enhance the risk of botulinum toxin production.

Since epidemiological surveys started in the early 1950s, there was a clear and positive trend in the number of botulism cases coincident with the declaration of Alaska statehood in 1958 and the introduction of Western culture and consumer goods. These figures reached their peak in the mid-1990s and have since been declining, a trend attributed to changing cultural tastes of younger generations.

It's not just the introduction of novel fermenting materials that is responsible for these outbreaks, but also the loss of indigenous knowledge regarding preservation techniques. Shaffer et al. described an outbreak of botulism caused by fermented salmon heads in 1985. The food preparer "had been taught to ferment foods by her mother, but had not prepared such foods in many years." Instead of placing the wooden barrel in the ground, she left the barrel aboveground exposed to the sun. The researchers noted that she retrospectively "recalled having been warned as a child that the sun's rays had a 'death meaning' and that fermented foods needed to be kept away from the 'killing rays of the sun.'"

Fermenting food is a delicate, complex process. As the Eskimo scholar Zona Spray noted in 2002, every step of the complex preparation process is carefully executed to ensure a highly acidic environment. She mentions that it is usually elders who prepare such traditional foods, and that they are better versed in the "oral history of health and sickness" than the younger generations.

This strongly suggests that a failure to transmit traditional knowledge and customs may play a pivotal role in the use of different preservation materials and in the skyrocketing incidence of botulism outbreaks in Alaska over the past fifty years. There's a quite grisly catalog compiled by C. E. Dolman in 1960 detailing botulinum deaths as a result of these improper storage practices; most cases seem to be variations on the theme of, for instance, eating rotted seal flipper stored in a gasoline barrel behind the stove. Death meaning, indeed!

This loss of wisdom in classical cooking knowledge is compounded by the introduction of novel gastronomical delights. Since the 1960s, fermented beaver paw and tail have recently contributed to the menu in southwestern Alaska. Beaver has been customarily boiled in stews, but the fermentation of this critter's extremities is an innovation. This nontraditional food item has since been incriminated in several botulism outbreaks.

In 2001, fourteen people fell ill after eating beaver tail and paw that had been fermented in a closed paper rice sack and stored for three months in a patient's house. Some of the meat hadn't fermented for very long, being added to the sack only within a week of its consumption. This change in dietetic patterns to include species that may be less amenable to the fermentation process than other customarily used meats also plays a role in some of the more recent botulism outbreaks.

So how do these *C. botulinum* spores find their way into my stink eggs and seal oil, you may ask? A number of studies have found high numbers of these unusually hardy spores in the soil and offshore waters along the Alaskan coast; these spores are quite resistant to the low temperatures of the North and can survive for months at a time. It is speculated that freshly caught fish and sea mammals butchered on beaches may be contaminated by soil, thereby introducing the bacterium to food destined for fermentation. This is apparently a common practice in Eskimo camps, where the carcasses of seals are piled on beach shores for days prior to being skinned and prepared for consumption. It is also been hypothesized that marine mammals and fish inadvertently ingest these spores during feeding, ultimately contributing to their intestinal flora; such "endogenous infection" may also serve as an origin of contaminated meat.

One may ask, why not just cook the meat to kill the spores and destroy the heat-labile toxin? And, in fact, both the spores and type E toxin are susceptible to high heats of 80°C for a short period of time. The possible response to such a suggestion is succinctly put by Dolman: "The proposed violation of time-honoured recipes may be rejected . . . because the desired flavour will be lost; it may be scorned in defiance of the white man's encroachments and paternalism; or ignored when fuel is scarce, time short, and hunger rampant."

Botulism poses a unique and recurring public health threat to Alaska Natives due to a pretty neat triumvirate of ecological, dietetic, and cultural factors that enhance the risk of infection. Modernized native lifestyles and loss of indigenous knowledge have clearly affected traditional methods of preparation and preservation, but fatal botulism outbreaks aren't a novel phenomenon strictly related to changing fermentation practices. There are several instances of Alaska Natives contracting the illness following consumption of the raw meat of long-since-deceased beached whales. In fact, whalers and Arctic explorers in the late 1890s and early 1900s describe entire Alaskan families dropping dead following consumption of semidecayed whale meat.

This is a dietary choice and culinary practice long informed by culture and history, which, in turn, are designated by some pretty profound ecological and climatic circumstances. I can't speak to the social and emotional ramifications of botulism infection, nor will I consider the consequences of the loss of indigenous culinary practices for younger generations. But think about it, would you? Botulism is a rare disease, and it's fascinating to find this niche where a preventable disease makes a strong appearance in the population due to cultural practices.

REFERENCES

Centers for Disease Control and Prevention. 2001. "Botulism Outbreak Associated with Eating Fermented Food—Alaska, 2001." *Morbidity and Mortality Weekly Report* 50 (32): 680–82. PMID: 11785568. www.cdc.gov/mmwr/preview/mmwrhtml/mm5032a2.htm.

Chiou, Lisa A., Thomas W. Hennessy, Andrea Horn, Gary Carter, and Jay C. Butler. 2002. "Botulism Among Alaska Natives in the Bristol Bay Area of Southwest Alaska: A Survey of Knowledge, Attitudes, and Practices Related to Fermented Foods Known to Cause Botulism." *International Journal of Circumpolar Health* 61 (1): 50–60.

Dolman, Claude E. 1960. "Type E Botulism: A Hazard of the North." *Arctic* 13 (4): 230–56.

Lancaster, Miriam J. 1990. "Botulism: North to Alaska." *American Journal of Nursing* 90 (1): 60–62.

McLaughlin, Joseph B., Jeremy Sobel, Tracey Lynn, Elizabeth Funk, and John P. Middaugh. 2004. "Botulism Type E Outbreak Associated with Eating a Beached Whale, Alaska." *Emerging Infectious Diseases* 10 (9): 1685–87.

Middaugh, John B., Tracey Lynn, B. Funk, and B. Jilly. 2003. "Outbreak of Botulism Type E Associated with Eating a Beached Whale—Western Alaska, July 2002." *Journal of the American Medical Association* 289 (7): 836–38. doi:10.1001/jama.289.7.836.

Miller, Lawrence G., Paul S. Clark, and George A. Kunkle. 1972. "Possible Origin of *Clostridium botulinum* Contamination of Eskimo Foods in Northwestern Alaska." *Applied Microbiology* 23 (2): 427–28.

Nelson, Edward W. 1900. *The Eskimo About Bering Strait.* Extract from the Eighteenth Annual Report of the Bureau of American Ethnology. Washington: United States Government Printing Office. http://openlibrary.org/books/OL7253164M/The_Eskimo_about_Bering_strait.

Shaffer, Nathan, Robert B. Wainwright, John P. Middaugh, and Robert V. Tauxe. 1990. "Botulism Among Alaska Natives: The Role of Changing Food Preparation and Consumption Practices." *Western Journal of Medicine* 153 (4): 390–93.

Spray, Zona. 2002. "Alaska's Vanishing Arctic Cuisine." *Gastronomica: The Journal of Food and Culture*. 2 (1): 30–40.

State of Alaska Department of Health and Social Services, Division of Public Health, Section of Epidemiology. 2005 update. "Botulism in Alaska: A Guide for Physicians and Healthcare Providers." www.epi.hss.state.ak.us/pubs/botulism/Botulism.pdf.

REBECCA KRESTON is an Aussie-American currently living in New Orleans. She is the founder of *Body Horrors* (http://bodyhorrors.wordpress.com), a blog exploring the anthropology, geography, and history of infectious diseases and parasites.

29

CHESSBOXING IS FIGHTING FOR GOOD BEHAVIOR

ANDREA KUSZEWSKI

Teleportation, cloaks of invisibility, smell-o-vision, 3-D printing, and even holograms were all ideas first imagined in science fiction—and now they are real products and technologies in various stages of development by scientists. While this is common in fields like experimental physics, it isn't as often that cognitive neuroscience and applied psychology score insights from this fantasy genre.

Chessboxing, a hybrid sport combining chess and boxing, made its first appearance in the pages of a 1992 sci-fi graphic novel by Enki Bilal, *Cold Equator* (part three of *The Nikopol Trilogy*). Combining what is described as the number one physical sport and the number one thinking sport into one completely new hybrid, chessboxing was meant to be the ultimate test of body and mind.

In 2003, Dutch artist Iepe Rubingh wanted to give that idea life. He saw the potential for an incredibly challenging new sport that would require physical strength and agility, superior problem solving skills, and above all, unbelievable mental discipline and control. No longer just a sci-fi fantasy, chessboxing is now one of the newest sport fads in Europe, quickly gaining popularity in the United Kingdom and the United States.

The most awesome thing about chessboxing—no, not the sci-fi roots or the extreme physical skill and mental prowess necessary for dominance—is the brain-changing potential of the sport itself. The nature of the execution of play, as well as the training involved, have some exciting implications for the future of aggression management and preventive treatment of maladaptive behaviors.

The ability to control aggression, emerging from a boxing ring?

This may seem unlikely, given that chessboxing is a contact sport, but let me explain.

Chessboxing is divided into eleven short, rapidly rotating rounds: six four-minute rounds of speed chess, alternated with five three-minute rounds of boxing. Winning is achieved by knockout (KO) or checkmate, or, in the case of a draw, points determine the winner.

The idea of performing at such a high physical level (boxing) as well as a high mental level (chess) seems arduous enough. But it isn't just the physical effort plus the mental effort of the two sports combined that makes it especially daunting. It's the constant alternating back and forth between the two that's the real challenge.

What is it about the alternating rounds that make this so intensely demanding? The answer lies largely in emotion regulation. The strength of a world-class boxer and the high rank of a chess player are of no use if a player lacks the one all-important skill: the ability to effectively regulate emotions in order to maintain cognitive control.

The Challenge of Task-Switching

Any chessboxer will tell you that the most difficult part of the sport is the moment you switch to the next round. Yes, the boxing is physically taxing, and chess takes extreme concentration, but it's the task-switching that poses the biggest challenge.

Here's why: During the boxing round, you need to anticipate your opponent's moves, plan your offensive/defensive strategy, exert physical force, while simultaneously blocking strikes. Naturally, you'll feel a rush of adrenaline. Your heart starts racing, and your emotional arousal spikes. After three minutes, however, when you hear that "ding" signifying the end of the round, you need to immediately let go of all of that emotion—and sit down to a four-minute round of chess. Fail to do this right away and you've set yourself up to be conquered. If you're too jazzed up from the boxing round, you won't be able to concentrate on your next chess move. So by the time you get in that chair facing the chessboard, you need to be ready to roll.

This takes a tremendous amount of cognitive control. You need to have intense focus during the boxing round, then let go of that and completely focus on the chess, then go back to complete focus on the boxing, all at lightning speed. The key to doing this without exhausting yourself to the point of physical and mental collapse is to keep your emotions low and controlled to begin with, never raising them above a certain threshold. If you are able to maintain a low level of emotional arousal, the task-switching is much easier, you can focus better on each separate task—less cognitive energy is spent down-regulating your emotions and more is spent on the executive functioning—and you maintain a better level of cognitive control. This puts you in a better position to defeat your opponent.

So why is emotion so important?

What Emotion Means for Behavior

Emotion regulation helps you to navigate your environment in order to meet the demands of any situation. This explains why having control over your emotions is so important in each individual task, and in maintenance of behavior overall. According to the book *Emotion Regulation and Psychopathology: A Transdiagnostic Approach to Etiology and Treatment* (2009): "Emotions function to interrupt ongoing cognitive processes or behavior, redirect attention to stimuli relevant to the preservation of goal-directed states, and trigger action tendencies in service of these goals."

In other words, your emotional state helps you to make sense of what is going on so you can choose the appropriate corresponding behaviors to engage in. It's that state of arousal—high, low, happy, sad—that will determine how your brain responds in individual situations. For example, if you're extremely nervous or upset, you will interpret an ambiguous situation much differently than you would if you were calm.

Additionally, if you are in a heightened arousal state (that is, too high for what that situation demands*), it's going to interfere with how well you are able to focus on a cognitive activity. Imagine what it would be like to get into a heated argument with someone right before you sit down to take an important exam. How do you think it would affect your performance?

*Research has shown that heightened arousal can help to motivate us to perform better at certain activities, but here we are speaking of a mismatch between how much arousal a person is experiencing and the optimal level of arousal for that specific circumstance.

The Benefit of Pre-Learning Emotional Control

If you've ever been in a situation like this, you know how difficult it is to focus when your emotions are through the roof. So now imagine that you were presented with the same situation, and instead of getting into an argument, you were able to maintain your cool, discuss the problem calmly, and never get yourself worked up emotionally. You would have a significantly easier time with that exam, given your calmer state.

Now what if I said we can actually train ourselves to maintain our cool when presented with emotionally charged situations—even before they happen?

Many behavior modification techniques involve practicing scenarios and either reinforcing the desired response, or using suppression or reappraisal in order to shape behavior in response to a problematic situation. This type of behavioral training is used during or after the targeted situation to help a person develop new behaviors to replace maladaptive ones, such as learning to stop and count to ten before responding in an argument.

These techniques have their successes, but researchers have found that when you shape the emotional response *before* the situation occurs, instead of modifying it afterward, there is a significantly higher success rate of change, and there are fewer negative side effects (such as increased heart rate and higher stress response). In other words, when the emotion regulation becomes automatic, the person doesn't feel as many negative effects as, say, someone who is actively trying to suppress a response.

To be in a stressful situation and have your emotions appropriately regulate automatically, prior to any negative response, is obviously the goal here. We want to be able to stand in the presence of adversity and take it on with a clear and purposeful mission, maintaining a level head. So how do we learn these Jedi mind tricks? And can chessboxing be that training medium to teach automatic emotion regulation?

Let's sum up what we now know about emotion regulation, put that together with some things we already know about behavior and learning in general, then tie it all in to chessboxing.

- When our emotions get too high, and we are unable to effectively down-regulate, cognition suffers. We may misinterpret situations, perform dismally, or make poor decisions.

- Emotion regulation that occurs automatically—a prior learned response to stressful situations—results in higher success at regulating behavior, with the fewest negative side effects.
- Recent research has shown that early exposure to mild stress—then working through strategies to manage that stress—can help to build not only resilience but better emotion regulation later in life, via strengthening of neural connections in the brain. This means if kids are taught from an early age to manage their emotional response to stress, they are more likely to become well-adjusted adults.
- This also means that it is possible to train your brain to better regulate emotions by engaging in activities that challenge you to maintain that low level of arousal. Over time, your ability improves to do this quickly and automatically.
- Research centered on treatment of psychological disorders has shown that teaching regulation of emotions to match the given environment can have a preventive effect on future displays of dysfunctional behavior. For example, if a person presents with manic episodes, teaching that person to regulate those extreme emotions (by the pre-learning, making-it-automatic method) can reduce the occurrence of those symptoms in the future. This is seen as a promising preventive treatment for mental disorders, or at least for some of their symptoms.
- The element of chessboxing that makes it so challenging is the task-switching—from adrenaline-pumping physical activity to intense cognitive activity, in short bursts—all while maintaining a low level of emotional arousal. Because you don't have the luxury of time to allow yourself to calm down naturally, you need to keep yourself from getting overly excited in the first place.
- Therefore, it follows that in order to be a good chessboxer, you need to be able to master automatic emotion regulation, and training in this sport will help to strengthen that skill.

The Exciting Part

Now that we have all of this information about the benefits of emotion regulation, how it can be trained and achieved? And what does this imply about future innovative methods of preventive treatment for problematic behavior?

Let's look at one example: anti-bullying strategies. As a behavior therapist working in schools, I've been witness to and participated in quite a few different types of anti-bullying programs. Each program varies in its exact content, but generally, they all seem to focus on how to react to bullying—either in reporting it to a teacher or in responding to the bully him/herself. These are all good things to teach. But what if we could target problem behavior before it even emerged?

Combination sports like chessboxing—ones that involve rapid switching between physical, adrenaline-producing activities and intense cognitive tasks—are ideal for training in emotion regulation. It may seem that an activity like boxing would promote aggression, but on the contrary, when combined with chess, utilizing short, alternating rounds, there is no time for aggression to build—otherwise you lose the game. If you train kids in how to control their emotional behavior or their aggression before they lash out at another child, then build on those patterns of successful behavior, you are one step ahead of the problem. Early intervention in any behavior problem almost always means a higher rate of success in the long term.

Granted, very young children probably aren't the ones who should be engaging in chessboxing specifically (the contact version is more appropriate for adolescents and young adults; there is a noncontact version of youth chessboxing), but any physical sport combined with alternating cognitive tasks, utilizing these same principles, would be just as great. Couldn't you see Kickball-Math, or Obstacle Course–Scavenger Hunts?

Rather than using a "Band-Aid approach" to address issues such as childhood aggression and bullying, let's use this scientific knowledge to come up with some innovative new solutions. By being a little creative, not only could we get kids enthusiastically involved in their own mental well-being, but such new mind-body sports might actually be a more effective way to treat behavior and aggression problems before they begin.

Is chessboxing the only solution for producing well-adjusted, emotionally stable adults? No, but it's a fun place to start. I, for one, am willing to fight a little for a better future. Are you?

REFERENCES

Calkins, Susan D. "Commentary: Conceptual and Methodological Challenges to the Study of Emotion Regulation and Psychopathology." *Journal of Psychopathology and Behavioral Assessment* 32 (2010): 92–95. doi:10.1007/s10862-009-9169-6.

Fairholme, Christopher P., Christina L. Boisseau, Kristen K. Ellard, Jill T. Ehrenreich, and David H. Barlow, "Emotions, Emotion Regulation, and Psychological Treatment: A Unified Perspective." In *Emotion Regulation and Psychopathology: A Transdiagnostic Approach to Etiology and Treatment*, edited by Ann M. Kring and Denise M. Sloan, 283–309. New York: Guilford Press, 2009.

Katz, Maor, Chunlei Liu, et al. "Prefrontal Plasticity and Stress Inoculation-Induced Resilience." *Developmental Neuroscience* 31, no. 4 (2009): 293–99.

Kring, Ann M. "The Future of Emotion Research in the Study of Psychopathology." *Emotion Review* 3, no. 2 (2010): 225–28.

Mauss, Iris B., Crystal L. Cook, and James J. Gross. "Automatic Emotion Regulation During Anger Provocation." *Journal of Experimental Social Psychology* 43 (2007): 698–711. doi:10.1016/j.jesp.2006.07.003.

Williams, Lawrence E., John A. Bargh, Christopher C. Nocera, and Jeremy R. Gray. "The Unconscious Regulation of Emotion: Nonconscious Reappraisal Goals Modulate Emotional Reactivity." *Emotion* 9, no. 6 (2009): 847–54.

World Chessboxing Organization. http://wcbo.org.

ANDREA KUSZEWSKI is a behavior therapist and consultant for children on the autism spectrum; her expertise is in Asperger's syndrome, or high-functioning autism. She works as a researcher with VORTEX: Integrative Science Improving Societies, studying creativity, and is also a fine artist herself, working with traditional drawing, digital painting, graphic design, and 3-D modeling and animation for the medical and behavioral sciences. She lives in Florida and blogs at *The Rogue Neuron*: www.science20.com/rogue_neuron.

30

MIRROR IMAGES: TWINS AND IDENTITY

DAVID MANLY

Whenever someone finds out I'm an identical twin, one question is always asked: "So, who's older?"

No matter which one of us is asked, my brother, Daniel, and I will always repeat the same answer we've given hundreds, if not thousands, of times before.

"Our parents never told us," says Daniel. "We always say that, along with the fact that they read in a book that by telling us which was older, it would give the older one a superiority complex and the younger child an inferiority complex. So we don't know."

Most identical twins know who was born first. In fact, as children, we were relentless in attempting to find out, but our parents refused. We asked in every way we could imagine, even pleaded with them to tell. But the only information we were able to extract was that the time difference between us was only five minutes, and that it really didn't matter.

After a long time, we both came to realize that our parents' decision helped us become the separate individuals that we are today. It allowed us to step out of each other's shadow and become David and Daniel—instead of the Manly twins.

If I was to find out that I was born first, or that Daniel was a few minutes older, I do not think it would affect my life in any big way. But to everyone else, it is always the first, and most important, question they ask.

"We're still us, after all," says Daniel. "That doesn't disappear. We are still the same people we always have been, regardless of who is older."

The Perils of Childhood

My parents were veterans of my "perfect" older sister, and thought they could handle their next child with relative ease. Suffice it to say, they were mistaken.

In fact, my parents still recall the first few years with a mix of horror and surprise.

"I cannot believe we survived," says my mom with awe. "We did not know what to expect with twins, but it was much harder than we thought."

They did read a lot of books on how to raise twins, and while the books did not help with actually raising us, they did help in other ways. These books helped our parents make some very important decisions early on that helped my brother and me become the individuals we are today, instead of simple carbon copies of the same person.

In addition to not divulging who was born first, our parents also split us into different classes in school as soon as possible. Their reasoning was that twins could easily fall into becoming the same person, and being 100 percent dependent on each other.

One of the most important things our parents ever did was instill into our minds that while we are twins, it should not be what defines us.

Identical Does Not Mean the Same

While most people can take solace in the fact that they are unique and one of a kind, twins do not have that. Because twins' DNA is practically identical when they are born, each must take a journey of self-discovery to forge his or her own identity, which is different than it is for those who are not twins (or as they are known in the twin community, "singletons").

Giving birth to identical, or monozygotic, twins is a relatively rare natural occurrence—it occurs in approximately 3.5 in every 1,000 births worldwide.[1] Yet in industrialized nations such as Canada and the United States, the total rate of twinning is now between 15 and 20 births per 1,000.[2] That's because the rate of fraternal (dizygotic) twins, and other multiple births, has risen sharply in the last fifty years, due primarily to increased maternal age and the widespread use of infertility treatments.

Dizygotic twins, and most larger multiple births, occur when two or more eggs are released and fertilized at the same time, which is greatly aided by the use of fertility treatments. Fraternal twins share approximately 50 percent of their DNA, like any other siblings. However, identical (or monozygotic) twins come from a single egg that splits into two separate embryos by a mechanism that is still not well understood, and they have almost 100 percent the same DNA.

Thus, each one of the thousands of twins born worldwide each day must struggle with establishing an identity that is unique and distinct from his or her genetic counterpart. For fraternal twins, who can even be different sexes, it is easier, as they are just siblings who happen to be the same age. But for identical twins, it is that much harder.

For example, Daniel and I are what is known as "mirror image twins," like approximately 25 percent of all twins. That means some of our features are the exact opposite of one another, such as our fingerprints.

Which is which? (Photograph courtesy of David Manly)

Imagine you had to craft a separate identity from someone who mirrored you exactly, with the same likes, dislikes, and so on . . .

How would you do it? I can tell you from personal experience that it is not easy.

Singletons do not know or appreciate how much time and effort is required for twins to step out of their previously established twin identity and for each to create a new one, all his or her own. In fact, making oneself an individual is one of the most difficult parts of being a twin. As G. Thomas Couser wrote bluntly in a 2003 article on the rarity of twins' autobiographies: "Twinhood can impede individuation."[3]

The Twin Stereotype

A major hurdle for a twin's developing individuality is an assumption that we all encounter from the general public—the twin stereotype.

For a twin, it is almost impossible not to run into these assumptions throughout life, such as the belief that we can read each other's thoughts, feel each other's pain, and are the exact same person. Not surprisingly, while many people believe there is some truth to these claims, they are all false.

Elise Milbradt, an identical twin, knows this all too well. She says that being a twin creates many complex obstacles that one must endure. "It's not just the constant confusion and the not knowing who is who—I mean, that's a part of it, but you get used to that. The worst part is that people really do believe that you are the same person."

My brother and I were, and still are, constantly mistaken for each other. Even our parents and our friends, some of whom have known us for almost twenty-eight years, still can and do confuse us from time to time.

This has led our family and friends to come up with a wide variety of ways to tell us apart, from the relatively simple to the just plain odd.

"I remember one girl in grade four," says Daniel, "who used to tell us apart by the color of our lunch boxes! That stuff is ridiculous, since the easiest way to tell us apart is to get to know us. Once you know one twin more than the other, you start to notice the differences . . . like I'm a little more reserved, while David is more outgoing. Or that David tends to talk a lot, while I usually just sit back and listen."

This constant confusion of who is who can create a rather difficult environment for individuality to develop in, which has led to many scientific studies on the subject.

The Science of Identity

One study done by Barbara Prainsack and Tim Spector in 2006 stated that, despite the strong bond between twins, the preconceptions of non-twins in their lives could potentially create a drive for one or both to seek out and claim a unique identity, distinct from that of their twin. The fact that twins are

> affected by the failure of some singletons to fully acknowledge the individuality of human beings who look very much alike and/or who go through large parts of their lives together, indicates that this is much rather a problem related to superficial characteristics of similarity (such as wearing similar clothes, having the same eye and hair color, etc.) than to identical genes.[4]

To overcome this problem, many parents of twins will attempt to manufacture some sort of a difference by dressing their children in a set color scheme—for example, I was always in blue and Daniel in red.

And it's not just us. Many parents choose separate color schemes for their twins.

For example, identical twin girls Amy and Jaclyn Jacobs were always dressed in yellow and pink, respectively.

According to Jaclyn, wearing those separate colors may have influenced her identity more than she thought. "It never occurred to me, but if you directly compare me and Amy, I'm the one who is more interested in fashion and girly things, while she doesn't really care about that stuff. Wearing pink all those years may have helped with that."

While many parents dress twins in similar clothes when they are babies and young children, that trend does not usually continue as they age. According to a study released in the U.K. in 2006, continuing to dress twins alike could result in the development of stigma and isolation from their peers.[5]

Kate Bacon, who authored the study, explored the dynamics of twinship with respect to twins' family environments and the development of their individual identities. She noted that one of the only ways for twins to be recognized as different is to work together to eliminate as much personality overlap as possible. "Whilst parents may 'set the stage' for twins' presentations of self, twins actively engage in the business of 'identity work' and utilize each other to try to manage other people's perceptions of them, to avoid social stigma and to bring off a convincing performance of who they are."

Bacon is referring to the strong bond that twins develop with one another, and her observation that in some cases, without each other they cannot truly express themselves. While I do not agree with all of her conclusions, I have noticed that I do not have to hide who I am from my brother.

This is because, as twins, we have spent much of our lives together and know each other very well. Jaclyn Jacobs says that this bond is closer than that with a friend, a sibling, or even a significant other.

Her sister, Amy, agrees. "Singletons cannot understand the bond between twins," she says. "It's like having someone around who understands the way you work."

My brother smiled when I told him this. "It's true," he says. "We just get each other. It's not telepathy or some sort of psychic connection—we've just been raised together, had similar experiences, and share many of the same likes and dislikes. Our brains just work in a similar way."

And yet, all the twins I spoke to, including my brother, agreed that there comes a time when being tethered to someone close to you is no longer needed.

"There came a point where I needed, really needed people to know me for me," says Elise Milbradt. "I wanted them to know me as Elise, the person who happened to be a twin, not the twin who happened to be named Elise.

"After all, you don't want to be twins forever."

And that is the real secret about twins. While we are twins, it is not the only thing that we want to be known for.

After all, everyone is an individual—even a twin.

NOTES

1. SOGC (Society of Obstetricians and Gynecologists of Canada), "Women's Health Information: Pregnancy: Multiple Birth," www.sogc.org/health/pregnancy-multiple_e.asp. Last updated September 1, 2010.

2. Chantal Hoekstra, Zhen Zhen Zhao, Cornelius B. Lambalk, Gonneke Willemsen, Nicholas G. Martin, Dorret I. Boomsma, and Grant W. Montgomery, "Dizygotic Twinning," *Human Reproduction Update* 14, no. 1 (2008): 37–47.
3. G. Thomas Couser, "Identity, Identicality, and Life Writing: Telling (the Silent) Twins Apart," *Biography* 26, no. 2 (Spring 2003): 243–60. doi:10.1353/bio.2003.0042.
4. Barbara Prainsack and Tim D. Spector, "Twins: A Cloning Experience," *Social Science and Medicine* 63 (2006): 2739–52. doi:10.1016/j.socscimed.2006.06.024.
5. Kate Bacon, "'It's Good to Be Different': Parent and Child Negotiations of 'Twin' Identity," *Twin Research and Human Genetics* 9, no. 1 (2006): 141–47. PMID: 16611479.

DAVID MANLY is a Canadian science journalist who holds degrees in biology and zoology, as well as a master's in journalism. He spends his time on Twitter (@davidmanly) discussing science communication, animals, and the general bizarre nature of day-to-day life. He and his twin brother, Daniel (a teacher), are very close, but still individuals. That said, since they are identical twins, they still do enjoy getting confused with each other from time to time.

31

DID THE CIA FAKE A VACCINATION CAMPAIGN?

MARYN McKENNA

A number of years ago, I was in New Delhi, at the end of an exhausting eighteen hours in which I had torn around the city to watch a National Immunization Day. On those days—like a national holiday, with flags and banners and kids let out from school—tens of millions of children line up to stick out their tongues and receive the sugary drops that contain the vaccine that should protect them against polio.

The Indian government, along with the Centers for Disease Control, the World Health Organization and the volunteer ground troops of Rotary International, has been organizing these days now for most of two decades, always coming closer to the goal of eradicating polio, never quite getting there.

On this day, which occurred close to the end of weeks I had spent embedded with a WHO STOP (Stop Transmission Of Polio) team, 135 million children were expected to queue in cities and suburbs and rich neighborhoods and slums. I spent the day with the team I had been observing, racing in a battered turquoise Tata from neighborhood to neighborhood, trying to understand where the campaign's message was working and where its earnest persuasions had failed.

There was one neighborhood, about fifteen miles outside the center of New Delhi, where things were not going well. It was a Muslim area, and the local masjids supported the campaign—all the imams had preached in favor of it—but the appeals had not penetrated. Only a few children, about twenty-five in a slum that held thousands, had wandered up to receive the drops and the swipe of gentian violet across a fingernail that would signal to canvassers that a child had been immunized.

(© Gary A. Korhonen)

A young mother selling fish in a muddy side street shrugged and dismissed the importance of her four children receiving the vaccine. In the next few days, she said, the government would send "mop-up" vaccinators into the alleys to find the children who had missed the big campaign. Maybe they would find her children; maybe she would get paying work that day, and then she would be away from home, and the children, too. The government's priorities were not hers.

Late that night, over whiskey sodas served without ice because of the water quality, a longtime local health worker explained what he thought was going on. Hypothetically, protecting against polio requires four rounds of drops. But in tropical temperatures, with inadequate sanitation and endemic diarrheal diseases, it can take many more rounds of immunization to ensure a child is immune. That unplanned-for reality had combined with the long-standing distrust between Hindus and Muslims to produce a situation that no one had foreseen.

"We come back to their neighborhoods, month after month, telling them that these drops will protect their children from being paralyzed," he said. "We come ten, eleven, twelve times and the kids become paralyzed anyway. They start to think we're doing this for some other reason, and they suspect us, and they don't want to bring the kids out anymore."

And that is why the CIA's decision to use a fake vaccination program in the hunt for Osama bin Laden, if that story is true, is such an appallingly, idiotically bad idea.

As reported by *The Guardian* on July 11, 2011,[1] and subsequently by *The New York Times*,[2] intelligence operatives funded a sham vaccination program in hopes of obtaining a sample of DNA to prove that bin Laden, then rumored to be in the area, was actually living in the compound where he was subsequently found and killed. From *The Guardian*:

> DNA from any of the Bin Laden children in the compound could be compared with a sample from his sister, who died in Boston in 2010, to provide evidence that the family was present.
>
> So agents approached [Shakil] Afridi, the health official in charge of Khyber, part of the tribal area that runs along the Afghan border.
>
> The doctor went to Abbottabad in March, saying he had procured funds to give free vaccinations for hepatitis B. Bypassing the management of the Abbottabad health services, he paid generous sums to low-ranking local government health workers, who took part in the operation without knowing about the connection to Bin Laden. Health visitors in the area were among the few people who had gained access to the Bin Laden compound in the past, administering polio drops to some of the children . . .
>
> In March health workers administered the vaccine in a poor neighbourhood on the edge of Abbottabad called Nawa Sher. The hepatitis B vaccine is usually given in three doses, the second a month after the first. But in April, instead of administering the second dose in Nawa Sher, the doctor returned to Abbottabad and moved the nurses on to Bilal Town, the suburb where Bin Laden lived.

There is no evidence the "vaccinations" produced DNA that helped identify bin Laden. The physician named in the article has been arrested by the Pakistani security forces. The CIA has understandably refused any comment. But the allegation that a vaccine program was not what it seemed—that it was not only suspect, but justifiably suspect—has been very widely reported.

This is awful. It plays, so precisely that it might have been scripted, into the most paranoid conspiracy theories about vaccines: that they are pointless, poisonous, covert shields for nefarious government agendas meant to do children harm.

That is not speculation. The polio campaign has already seen this happen, based on just those kind of suspicions—not just in a single poor slum in New Delhi, but across much of Sub-Saharan Africa.

In the fall of 2003, a group of imams in the northern Nigerian state of Kano—the area that happened to have the highest rate of ongoing polio transmission—began preaching against polio vaccination, contending that what purported to be a protective act was actually a covert campaign by Western powers to sterilize and kill Muslim children. The president of Nigeria's Supreme Council for Sharia Law said to the BBC: "There were strong reasons to believe that the polio immunisation vaccine was contaminated with anti-fertility drugs, contaminated with certain virus that cause HIV/AIDS, contaminated with Simian virus that are likely to cause cancers."

The rumors caught like wildfire, and they were spread further by political operatives who saw an opportunity to disrupt a recent post-election power-sharing agreement between the Muslim north and the Christian south. Three majority-Muslim states—Kano, Kaduna, and Zamfara—suspended polio vaccination entirely. Vaccination acceptance in the rest of the country fell off so sharply that the national government was forced to act. It ordered tests of the vaccine by Nigeria's health ministry and empaneled a special commission to visit the Indonesian labs where the vaccine administered in Nigeria was made. The WHO convened emergency meetings.

And polio began to spread. At the end of 2003, when the boycott began, there had been only 784 known polio cases in the entire world. By the end of 2004, there had been 793 new cases just in Nigeria. Polio leaking across Nigeria's borders reinfected Benin, Botswana, Burkina Faso, Cameroon, Chad, the Central African Republic, Cote d'Ivoire, Ethiopia, Ghana, Guinea, Mali, Sudan, and Togo. Nigerian strains appeared in Yemen, site of the largest port on the Red Sea, and in Saudi

Arabia, imperiling the millions of pilgrims coming to the country on hajj.

The last holdouts in Kano did not fully accept polio vaccination until the end of 2004, leaving children born in 2003 unprotected. In 2006, the lab-weakened virus used in the vaccine underwent a known but extremely rare mutation back to the virulence of wild poliovirus. It caused an epidemic that ripped through the unprotected children—and simultaneously strengthened suspicions that the vaccine had always been dangerous.

Though the accusations that polio vaccination was a Potemkin cover for anti-Islamic activities were false, they almost ruined the international eradication campaign. Those accusations could be leveled again now, thanks to the CIA's alleged appalling ruse in Pakistan. Because of the ruse, the accusations will seem much more believable—and might even be true.

Longtime global-health reporter Tom Paulson says this will "undermine global health,"[5] and expands on the possible consequences; Seth Mnookin, author of *The Panic Virus,* calls it a "horrible move with potentially dangerous consequences";[6] infectious-disease physician Kent Sepkowitz says it's a "paranoid's dreamy nightmare";[7] and blogger Brett Keller bluntly calls it "despicable."[8] *The Guardian*'s Sarah Boseley adds: "a black day for medical ethics and a one-off crazy scheme."[9] And James Fallows at *The Atlantic* warns this has "tremendously damaging implications that must be addressed."[10]

I agree with them all.

Addenda: The misrepresented campaign remained a live news story for weeks; I kept track of developments via updates added to the end of the original post.

Update 1 [July 13, 2011]: And here we go. The Associated Press reported this afternoon:

> Pakistani health officials held meetings about the alleged CIA scheme on Tuesday and expressed concern that it could have a negative impact on immunization programs in other areas of

> the northwest, especially in Pakistan's semiautonomous tribal region along the Afghan border, said a Pakistani official involved in polio eradication efforts . . .
>
> One of the Pakistani Taliban's top commanders, Maulvi Faqir Mohammed, recently called on people in the northwest to avoid vaccines offered by the international community, claiming they were made with "extracts from bones and fat of an animal prohibited by God—the pig."
>
> "Don't fall prey to these infidel NGOs and this U.S.-allied government and its army," said Mohammed over the illegal radio station he transmits from his sanctuary in eastern Afghanistan.
>
> Pakistani officials and their international partners have pushed back against these claims, but the CIA's reported activities in the country may have made their job that much harder. "The medical mission has to be immune from manipulation for political and military purposes and health care workers generally must not be compelled to conduct activities contrary to medical ethics," said [Michael O'Brien, a spokesman for the International Committee of the Red Cross in Pakistan].[11]

Update 2 [July 13, 2011]: Wednesday evening, U.S. sources acknowledged that the vaccine campaign did happen. They insisted that real hepatitis vaccine was used. They did not address the fact that, since only one of three doses was delivered, the vaccination was effectively useless. From *The Washington Post*:

> U.S. officials on Wednesday defended a tactic used by the CIA to attempt to verify the whereabouts of Osama bin Laden—the covert creation of a vaccine program in Abbottabad, the town in Pakistan where he was later killed in a U.S. raid . . .
>
> A senior U.S. official said the campaign involved actual hepatitis vaccine and should not be construed as a "fake public health effort."
>
> "People need to put this into some perspective," said the official, speaking on the condition of anonymity because of the sensitivity of the issue. "The vaccination campaign was part of the hunt for the world's top terrorist, and nothing else. If the

United States hadn't shown this kind of creativity, people would be scratching their heads asking why it hadn't used all tools at its disposal to find bin Laden."

. . . The senior U.S. official declined to say whether DNA from bin Laden's relatives was collected as part of the vaccine program.[12]

Update 3 [July 14, 2011]: Thursday, Doctors without Borders (known in most of the world as MSF, for Médecins Sans Frontières), which has teams in a number of Pakistani and border provinces, released a statement saying the fake campaign interferes with health care and endangers health workers:

> "The mere suggestion that the provision of medical care was carried out under false pretenses damages public perception of the true purpose of medical action," said Dr. Unni Karunakara, MSF's international president. "With all populations in crisis, it is challenging enough for health agencies and humanitarian aid workers to gain access to, and the trust of, communities—especially populations already skeptical of the motives of any outside assistance . . . The risk is that vulnerable communities—anywhere—needing access to essential health services will understandably question the true motivation of medical workers and humanitarian aid. The potential consequence is that even basic health care, including vaccination, does not reach those who need it most."[13]

NOTES

1. Saeed Shah, "CIA Organised Fake Vaccination Drive to Get Osama bin Laden's Family DNA," *Guardian*, July 11, 2011. www.guardian.co.uk/world/2011/jul/11/cia-fake-vaccinations-osama-bin-ladens-dna.
2. Mark Mazzetti, "Vaccination Ruse Used in Pursuit of bin Laden," *New York Times*, July 11, 2011. www.nytimes.com/2011/07/12/world/asia/12dna.html.
3. BBC News, "Polio in Nigeria Threatens Region," October 27, 2003. http://news.bbc.co.uk/2/hi/africa/3216329.stm.
4. Laura MacInnis, "Nigeria Fights Rare Vaccine-Derived Polio Outbreak," Reuters,

October, 9, 2007. http://in.reuters.com/article/2007/10/08/us-nigeria-polio-idINLAU88516220071008.

5. Tom Paulson, "Did CIA Undermine Global Health by Faking Vaccines in Hunt for bin Laden?" Humanosphere, KPLU, Seattle, July 12, 2011. humanosphere.kplu.org/2011/07/has-the-cia-compromised-global-health-in-hunt-for-bin-laden/.
6. Seth Mnookin, "What Price Will the Black-Ops bin Laden Vaccination Scheme Exact on Global Health Efforts?" *The Panic Virus* (PLoS blog), July 12, 2011. http://blogs.plos.org/thepanicvirus/2011/07/12/what-price-will-the-black-ops-bin-laden-vaccination-scheme-exact-on-global-health-efforts/.
7. Kent Sepkowitz, M.D., "The CIA's Dangerous Vaccine Stunt," *Daily Beast*, July 12, 2011. www.thedailybeast.com/articles/2011/07/12/cia-s-bin-laden-vaccine-ruse-a-public-health-nightmare.html.
8. Brett Keller, "CIA's Despicable Pakistan Vaccination Ploy," *Brett Keller: Public Health & Development* (blog), July 11, 2011. www.bdkeller.com/2011/07/cias-despicable-pakistan-vaccination-ploy/.
9. Sarah Boseley, "Collateral Damage from the Hunt for bin Laden," *Sarah Boseley's Global Health Blog, Guardian,* July 13, 2011. www.guardian.co.uk/society/sarah-boseley-global-health/2011/jul/13/cia-vaccines.
10. James Fallows, "Did the CIA Really Use Fake Vaccinations to Get bin Laden?" *The Atlantic,* July 13, 2011. www.theatlantic.com/international/archive/2011/07/did-the-cia-really-use-fake-vaccinations-to-get-bin-laden/241863/.
11. Sebastian Abbott, "Reported CIA Vaccine Ruse Sparks Fear in Pakistan," Associated Press, July 13, 2011. http://news.yahoo.com/reported-cia-vaccine-ruse-sparks-fear-pakistan-165110692.html.
12. Jason Ukman, "CIA Defends Running Vaccine Program to Find bin Laden," *Washington Post,* July 13, 2011. www.washingtonpost.com/world/national-security/cia-defends-running-vaccine-program-to-find-bin-laden/2011/07/13/gIQAbLcFDI_story.html.
13. Médecins Sans Frontières, "Alleged Fake CIA Vaccination Campaign Undermines Medical Care," press release, July 14, 2011. www.doctorswithoutborders.org/press/release.cfm?id=5439&cat=press-release.

MARYN MCKENNA is a blogger for Wired.com, a columnist for *Scientific American,* a magazine journalist, and the author of *Superbug: The Fatal Menace of* MRSA and *Beating Back the Devil: On the Front Lines with the Disease Detectives of the Epidemic Intelligence Service.*

32

FREE WILL AND QUANTUM CLONES

GEORGE MUSSER JR.

The late philosopher Robert Nozick, talking about the deep question of why there is something rather than nothing, quipped: "Someone who proposes a non-strange answer shows he didn't understand the question."

So, when Scott Aaronson began a talk by saying it would be "the looniest talk I've ever given," it was a good start. At a conference on the nature of time—a question so deep it's hard even to formulate as a question—"loony" is high praise indeed. And indeed his talk was rich in ambition and vision. It left physics überblogger Sabine Hossenfelder uncharacteristically lost for words.

As part of his general push to apply theoretical computer science to philosophy, Aaronson has been giving thought to that old favorite of college metaphysics classes and late-night dorm-room bull sessions: free will. Do we have autonomy, or are our choices preordained? Is that a false choice? What does it mean to be free, anyway? Though hard to summarize, his talk can be broken down into two parts.

First, he sought to translate fuzzy notions of free will into a concrete operational definition. He proposed a variation on the Turing Test that he calls the Envelope Argument or Prediction Game: someone poses questions to you and to a computer model of your brain, trying to figure out who's the human. If a computer, operating deterministically, can reproduce your answers, then you, too, must be operating deterministically and are therefore not truly free. (Here I use the word "deterministically" in a physicist's or philosopher's sense; computer scientists have their own, narrower meaning.) Although the test can never be definitive, the unpredictability of your responses can be quantified by the size of the smallest computer program needed to reproduce those responses.

The output of this game, as Aaronson portrayed it, would be a level of confidence for whether your will is free or not. But I think it might be better interpreted as a measure of the amount of free will you have. In 2010, quantum physicists Jonathan Barrett and Nicolas Gisin argued that free will is not a binary choice, live free or die, but a power that admits of degree. They proposed to quantify free will using quantum entanglement experiments. Freedom of will enters into these experiments because physicists make a choice about which property of a particle to measure, and the choice affects the outcome. Such experiments are commonly taken as evidence for spooky action at a distance, because your choice can affect the outcome of a measurement made at a distant location. But they can also be interpreted as a probe of free will.

If there are, say, 1,000 possible measurements, then complete freedom means you could choose any of the 1,000; if your choice were constrained to 500, you would have lost one bit of free will. Barrett and Gisin showed that the loss of even a single bit would explain away spooky action. You wouldn't need to suppose that your decision somehow leaps across space to influence the particle. Instead, both your choice and the outcome could be prearranged to match.

What is surprising is how little advance setup would do the trick. The more you think about this, the more disturbed you should get. Science experiments always presume complete freedom of will; without it, how would we know that some grand conspiracy isn't manipulating our choices to hide the truth from us?

Back to Aaronson. After describing his experiment, he posed the question of whether a computer could ever convincingly win the Prediction Game. The trouble is that a crucial step—doing a brain scan to set up the computer model—cannot be done with fidelity. Quantum mechanics forbids you from making a perfect copy of a quantum state—a principle known as the no-cloning theorem.

The significance of this depends on how strongly quantum effects operate in the brain. If the mind is mostly classical, then the computer could predict most of your decisions.

Invoking the no-cloning theorem is a clever twist. The theorem derives from the determinism—technically, unitarity—of quantum mechanics. So here we have determinism acting not as the slayer of free will, but as its savior. Quantum mechanics is a theory with a keen sense of irony. In the process of quantum decoherence, to give another example, entanglement is destroyed by . . . more entanglement.

As fun as Aaronson's game is, I don't see it as a test of free will per se. As he admitted, predictable does not mean unfree. Predictability is just one aspect of the problem. In the spirit of inventing variations on the Turing Test, consider the Toddler Test. Ask a toddler something, anything. He or she will say no. It is a test that parents will wearily recognize. The answers, by Aaronson's complexity measure, are completely predictable. But that hardly reflects on the toddler's freedom; indeed, toddlers play the game precisely to exercise their free will. The Toddler Test shows the limits of predictability, too. Who knows when the toddler will stop playing? If there is anybody in the world who is unpredictable, it is a toddler. What parents would give for a window in their skulls!

Yet no one denies that toddlers are composed of particles that behave according to deterministic laws. So how do you square their free will with those laws? Like cosmologist Sean Carroll, I lean toward what philosophers call compatibilism: I see no contradiction whatsoever between determinism and free will, because they operate at two different levels of reality. Determinism describes the basic laws of physics. Free will describes the behavior of conscious beings. It is an emergent property. Individual particles aren't free. Nor are they hot or wet or alive. Those properties arise from particles' collective behavior.

To put it differently, we can't talk about whether you have free will until we can talk about *you*. The behavior of particles could be completely preordained by the initial conditions of the universe, but that is irrelevant to your decisions. You still need to make them.

What you are is the confluence of countless chains of events that stretch back to the dawn of time. Every decision you make depends on everything you have ever learned and experienced, coming together in your head for the first and only time in the history of the universe. The decision you make is implicit in those influences, but they have never all intersected before. Thus your decision is a unique creative act.

This is why even the slightest violation of free will in a quantum entanglement experiment beggars belief. "Free will" in such an experiment means simply that your choice of what to measure is such a distant cousin of the particle's behavior that the two have never interacted until now.

This is where we get into the second big point. Even if the influences producing a free choice have never interacted before, they can all be traced to the initial state of the universe. There is always some uncertainty about what that state was; a huge range of possibilities would have led to the universe we see today. But the decision you make resolves some of that uncertainty. It acts as a measurement of those countless influences.

Yet in a deterministic universe, there is no justification for saying that the initial state caused the decision; it is equally valid to say that the decision caused the initial state. After all, physics is reversible. What determinism means is that the state at one time implies the state at all other times. It does not privilege one state over another. Thus your decision, in a very real sense, creates the initial conditions of the universe.

This backward causation, or retrocausality, was the "loony" aspect of Aaronson's talk. Except there's nothing loony about it. It is a concept that Einstein's special theory of relativity made a live possibility. Relativity convinced most physicists that we live in a "block universe" in which past, present, and future are equally real. In that case, there's no reason to suppose the past influences the future, but not vice versa. Although their theories shout retrocausality, physicists haven't fully grappled with the implications yet. It might, for one thing, explain many of the mysteries of quantum mechanics.

In a follow-up e-mail, Aaronson told me that the connection between free will and the cosmic initial state was also explored by philosopher Carl Hoefer in a 2002 paper. What Aaronson has done is apply the insights of quantum mechanics. If you can't clone a quantum state perfectly, you can't clone yourself perfectly, and if you can't clone yourself perfectly, you can't ever be fully simulated on a computer. Each decision you take is yours and yours alone. It is the unique record of some far-flung collection of particles in the early universe.

Aaronson wrote, "What quantum mechanics lets you do here, basically, is ensure that the aspects of the initial microstate that are getting resolved with each decision are 'fresh' aspects, which haven't been measured or recorded by anyone else."

If nothing else, let this reconcile parents to their willful toddlers. Carroll once wrote that every time you break an egg, you are doing observational cosmology. A toddler playing the "no" game goes you one better. Every time the toddler says no, he or she is doing cosmological engineering, helping to shape the initial state of the universe.

GEORGE MUSSER JR. is a senior editor, writer, and general troublemaker at *Scientific American*. He is the author of *The Complete Idiot's Guide to String Theory* (2008) and winner of the 2011 Science Writing Award from the American Institute of Physics. His website is www.buckyspace.com.

33

DEAR EMMA B

PZ MYERS

Ken Ham, the president of the young Earth creationist organization Answers in Genesis (AiG), is crowing over fooling a child. A young girl visited a Moon rock display from NASA, and bravely went up to the docent and asked the standard question Ham coaches kids to ask—and she's quite proud of herself. Ham quotes her letter on his blog:

> I went to a NASA display of a moon rock and a lady said, "This Moon-rock is 3.75 billion years old!" Guess what I asked for the first time ever?
>
> "Um, may I ask a question?"
>
> And she said, "Of course."
>
> I said, in my most polite voice, "Were you there?"
>
> Love, Emma B

Ken Ham is also quite proud of himself. He's pleased with the fact that many people will be dismayed at the miseducation he delivers:

> Each time I give examples in my blog posts of children who have been influenced by AiG, the atheists go ballistic on their blogs. They hate to read of instances like this. They want to teach these children there is no God and they are just animals in this hopeless and meaningless struggle of this purposeless existence.

I am angry at Ken Ham, but in this case, I mainly feel sad for Emma B, who is being manipulated and harmed by a delusion. So I thought what I would do is write a letter to her—a letter I wouldn't send, because I'm not going to intrude on a family with the actual science, but because this is what I would say if Emma actually asked me.

Dear Emma:

I read your account of seeing a 3.75-billion-year-old Moon rock, and how you asked the person displaying it "Were you there?"—the question that Ken Ham taught you to ask scientists. I'm glad you were asking questions—that's what scientists are supposed to do—but I have to explain to you that that wasn't a very good question, and that Ken Ham is a poor teacher. There are better questions you could have asked.

One serious problem with the "Were you there?" question is that it is not very sincere. You knew the answer already! You knew that woman had not been to the Moon, and you definitely knew that she had not been around to see the rock forming 3.75 billion years ago. You knew the only answer she could give was "no," which is not very informative.

Another problem is that if we can trust only what we have seen with our own two eyes in our short lives, then there's very little we can know at all. You probably know that there are penguins in Antarctica, and that the Civil War was fought in the 1860s, and that there are fish swimming deep in the ocean, and you also believe that Jesus was crucified two thousand years ago, but if I asked you "Were you there?" about each of those facts, you'd also have to answer "no" to each one. Does that mean they are all false?

Of course not. You know those things because you have other kinds of evidence. There are photographs and movies of penguins and fish; there are documents from the time of the Civil War, as well as the fact that in many places you can still find old bullets and cannonballs buried in the ground from the time of the war; and you have a book, the Bible, that tells stories about Jesus. You have evidence other than that you personally witnessed something.

This is important because we live in a big ol' beautiful world, far older than your nine years, and there's so much to learn about it—far more than you'll ever be able to see for yourself. There's a gigantic universe beyond South Carolina, and while you probably won't ever visit a distant star or go inside a cell, there are instruments we can use to see farther and deeper than your eyes can go, and there are books that describe all kinds of wonders. Don't close yourself off to them simply because you weren't there.

I'd like to teach you a different easy question, one that is far, far more useful than Ken Ham's silly "Were you there?" The question you can always ask is "How do you know that?"

Right away, you should be able to see the difference. You already knew the answer to the "Were you there?" question, but you don't know the answer to the "How do you know that?" question. That means the person answering it will tell you something you don't know, and you will learn something new. And that is the coolest thing ever.

You could have asked the lady at the exhibit, "How do you know that Moon rock is 3.75 billion years old?" and she would have explained it to you. Maybe you would disagree with her; maybe you'd think there's a better answer; maybe you'd still want to believe Ken Ham, who is not a scientist; but the important thing is that you'd have learned why she thought the rock was that old, and why scientists have said that it is that old, and how they worked out the age, even if they weren't there. And you'd be a little bit more knowledgeable today.

I'll assume you're actually interested in knowing how they figured out the age of the rock, so I'll try to explain it to you.

The technique scientists use is called radiometric dating. It uses the fact that some radioactive elements slowly fall apart, turning into other elements. For instance, a radioactive isotope of potassium will decay over time into an isotope of another element, argon.

One way to think of it is that it's like an hourglass. You know how they work: you start with all the sand in the top half of the hourglass, and it slowly trickles into the bottom half. If you see an hourglass with all the sand at the top and none at the bottom, you know it was recently flipped over. If you see one with half the sand in the top, and half in the bottom, you know it's about halfway through the time it will run. And if you look at how quickly the sand moves through the neck of the hourglass, you could even figure out how long until it all runs out.

In radiometric dating, the scientists are looking at how far along all the radioactive potassium is in the process of turning into argon. The amount of potassium is like the amount of sand in the top half of the hourglass, while the amount of argon is like the amount in the bottom half. By measuring the relative amounts of the two elements, and by measuring how fast radioactive potassium turns into argon, we can figure out how long it's been since the rock solidified.

It takes a very long time for the decay to occur. It takes 1.25 billion years for half the potassium to turn into argon. When they measured those elements in the Moon rocks, they found that the radiometric hourglass had mostly run out, so they knew that it was very, very old.

Scientists double-check everything. They also looked at other elements, like how quickly uranium turns into lead, or rubidium into strontium, and they all agree on the date, even though these are decay processes that run at different rates. All the radiometric hourglasses they've used give the same answer: 3.75 billion years. None of them say 6,000 years.

I think you're off to a great start—being brave enough to ask older people to explain themselves is exactly what you need to do to learn more and more, and open up the whole new exciting world of science for yourself. But that means you have to ask good questions to get good answers so that you will learn more.

Don't use Ken Ham's bad question, and most important, don't pay attention to Ken Ham's bad answers. There's a wealth of wonderful truths that reveal so much more about our universe out there, and you do not want to close your eyes to them. Maybe someday you could be a woman who does go to the Moon and sees the rocks there, or a geologist who sees how rocks erode and form here on Earth, or the biologist who observes life in exotic parts of the world . . . but you won't achieve any of those things if you limit your mind to the dogma of Answers in Genesis.

Best wishes for future learning,

PZ Myers

PZ MYERS is a biologist who teaches at the University of Minnesota Morris and blogs online at http://pharyngula.org.

34

FASTER THAN A SPEEDING PHOTON

CHAD ORZEL

In September, the OPERA collaboration at the Gran Sasso laboratory in Italy made the most dramatic physics announcement of the year, reporting that they had observed neutrinos (a fundamental particle with tiny mass and no charge) moving at speeds slightly greater than the speed of light in vacuum. Einstein's special theory of relativity says that no particle with mass can move faster than light, so this result, if it holds up, would overturn more than a century of physics, requiring modern physics to be rewritten from the ground up.

As you would expect in the wake of such an announcement, a lot of pixels were spilled over this faster-than-light neutrino business. It was the biggest physics story of the year, on blogs and in mass media, but too much of the commentary was of the form "*I am a [theorist, journalist], so hearing about experimental details gives me the vapors*" (a snarky paraphrase, obviously). This left room for a write-up at the level of the physics conversations I have with my dog, going into a bit more depth about what the researchers did and where it might be wrong.

So, what did those jokers at CERN pull this time? Isn't it bad enough that they want to feed us all into a black hole? Now they're messing with the speed of light?

First of all, this wasn't a CERN experiment in the same way that the Large Hadron Collider (LHC) is. The experiment uses a particle accelerator at CERN, but it's actually an Italian collaboration that did the experiment and data analysis, with the principal detector based at the Gran Sasso underground laboratory in the Alps. The name of the collaboration is OPERA, which is one of those ghastly pseudo-acronyms where they use the second letter of one of the words in order to force it to spell something.

Okay, fine. What have the Italians done?

Well, the goal of their experiment is to look for an "oscillation" of neutrinos. The neutrinos are created at CERN in one of their three varieties, and on the way to Gran Sasso, they can change character and end up being detected as a different type. They start as muon neutrinos and end up as tau neutrinos, or at least that's the plan.

As part of the preliminary analysis for their main experiment, the scientists looked at about three years' worth of data, and noticed something odd: the neutrinos in their experiment seem to be moving slightly faster than the speed of light. The difference is pretty big in absolute terms—about 7,500 m/s, or nearly 17,000 mph—but it's only about 1/40,000th of the speed of light. Still, the difference they see is many times larger than their uncertainty, and they can't figure out why, so they're making their results public.

Wow. How do they measure that, anyway?

Conceptually, what they did is the most basic kind of velocity measurement, the sort of thing we talk about in the first few weeks of introductory physics. They measured the distance between CERN and Gran Sasso, which is 730 kilometers, or 453.6 miles, and divided that by the time between when the neutrinos are created and when they're detected to get the speed at which the neutrinos covered that distance.

The implementation, of course, is a little more complicated . . .

But if they know when a neutrino was created and when it was detected, how could they screw this up?

That's the first complication. Neutrinos are ridiculously difficult to detect, so they have to throw a whole lot of them at their detector to see anything. In the entire three-year data set they analyzed, they have only around 16,000 neutrinos total.

This means that they don't actually have the ability to say when a specific neutrino was created, because there are huge numbers of neutrinos created for every one that they detect.

How can they possibly claim anything about the speed, then?

What they do is to compare the distribution of times when they detect neutrinos to the distribution of times when neutrinos were created in the source. The way this works is that they blast a moderately high-energy proton beam into a stack of graphite, which creates some exotic particles that eventually decay into other things, spitting out neutrinos along the way. The other decay products get blocked by the several hundred kilometers of rock between CERN and Gran Sasso, but the neutrinos interact so weakly with ordinary matter that they go right through, and a handful of them get detected.

They can't tell much of anything about exactly when the neutrinos are created, but they know that the neutrinos had to come from one of the protons in their original beam. And they have very good diagnostics on the proton beam. So they argue that the distribution of arrival times of the neutrinos ought to have the same shape as the distribution of arrival times of the protons. They compare those distributions as a whole in Figure 11 of their preprint on arXiv (see the URL at the end of this article). They find that a graph of the number of neutrinos detected in a particular 150-nanosecond interval near the time of a beam pulse looks very much like a similar graph for the number of protons hitting the graphite target, only shifted a bit in time. When they apply a correction for the travel time (basically, subtracting a microsecond from the time value for all the neutrino points), the two distributions line up beautifully.

So the time they subtract off is the thing they call the time of flight?

Exactly. If you look at the time that gives the best alignment between the two distributions and compare it to what you expect for particles moving at the speed of light, you find that it's about 60 ns too short, suggesting the neutrinos were moving faster than the speed of light.

Following the furor over their initial announcement, they repeated the experiment, in October and early November, using very short proton pulses, so they have a better idea when each individual neutrino was detected (to within a few nanoseconds), and while they detected only about twenty neutrinos in that short time, the time of flight that they measured agrees with their original measurement. This shows that if there is a flaw in their experiment, it's not from the comparison between distributions.

Okay, so how did they get the time wrong?

Well, that's the problem. They think they can account for all the uncertainties in the timing process, and it's not enough to cover the observed shortfall. In fact, it's about six times bigger than they think their measurement uncertainty is.

How do you get that kind of timing accuracy, anyway? I mean, I can't buy a watch that's good to better than a second. How do you keep track of things to the nanosecond?

Nanosecond timing isn't all that difficult by itself—you can get an atomic clock easily enough. What's tricky is synchronizing the timing at two different places. That's a fiendishly difficult problem, and it's more or less that sort of thing that led to the whole theory of relativity.

How do they do the synchronization?

They take advantage of GPS for that. The Global Positioning System is a network of satellites containing atomic clocks, each beaming out a signal saying what time it is according to that clock. Measuring the difference in times between several satellites lets you figure out how far you are from each of them, and since the orbits of the satellites are well known, that tells you where you are on the surface of Earth.

They use a so-called common-view GPS system to coordinate their timing. They have an atomic clock at each end of the experiment, and as the name suggests, they use the time broadcast by GPS satellites that are visible to both ends at the same time as a reference to make sure their clocks are synched up. The difference between the two sites is around two nanoseconds, much smaller than the arrival time shift they see.

So, is that the only problem with the timing?

Not quite. There's also some ambiguity about where the neutrinos are created, which might cause a problem. The exotic particles created by the proton beam pass down a one-kilometer tunnel, a shade over six-tenths of a mile, and the decay that produces the neutrinos happens somewhere in there, but they can't tell exactly where.

Wait a minute—their source is somewhere in a mysterious tunnel, but that doesn't throw off their timing accuracy?

The claim is that the particles that decay to make the neutrinos are also moving at very nearly the speed of light, and so pass down the tunnel at almost the same speed as the neutrinos. This makes the measurement less sensitive than you might think as to exactly where the neutrinos are created. What matters is not the total uncertainty in the position, but the difference in travel time for a neutrino created partway up the pipe versus one created all the way at the end; and the speed of the parent particles is so close to the speed of the neutrinos that there's very little difference in travel time. They've checked this with simulations (a standard practice in experimental particle physics), and say it's a tiny effect—a fraction of a nanosecond.

Okay, that's two places the timing might be wrong, but they say it isn't. How about the distance? Could they have screwed that up?

That's the other obvious source of error, but it's hard to see how. Again, they have GPS to use for this, and while the accuracy of the position obtained by GPS for a moving receiver, like in your phone, is only several yards, if you're trying to measure the distance between two fixed points, and monitor it over a long time, you can get really good accuracy. They claim to have the distance down to twenty centimeters, almost eight inches, which is a bit less than a nanosecond at the speed of light.

Twenty centimeters? Really?

Really. They even provide a graph showing their measurements over the three-year run, which pick up a slow change in the distance due to continental drift, and a dramatic jump due to an earthquake in 2009.

You know what? Living in the future is pretty awesome!

Yes, yes it is.

Okay, so, what—did the distance shrink in the winter or something?

It doesn't look that way. They looked for changes in the arrival time for different times of day, different times of year, and so on, and didn't find any significant difference. They've also checked the distance by other methods—sending light down fiber optics, etc.—and their answers agree. As far as they can tell, they know the distance to within 20 centimeters out of 730 km, full stop.

So, what else could be wrong?

Well, that's the problem. They've checked all the obvious things, and they all seem to hang together. Which is why they're putting this result out there, knowing full well that it disagrees with just about everything else. They're hoping that some clever person will spot a mistake, or, failing that, that other experimenters will do the same test (there's one experiment in Japan and one in the United States) and see if they get the same result.

Doesn't this conflict with the supernova measurements Ethan Siegel was going on about on his blog?

Probably, but maybe not. The measurement Ethan wrote up was a pulse of neutrinos detected from Supernova 1987A, in the Large Magellanic Cloud. The neutrinos arrived at almost the same time as the light from the supernova, giving a maximum speed for the neutrinos that is significantly lower than the speed the OPERA result suggests.

The neutrinos involved there had very low energies, though, compared to the ones used in the OPERA beam. It could be that really high-energy neutrinos travel at a different speed than really low-energy ones. They don't see any such dependence on energy in their results, but they can't check that much of an energy range, so it might be that there's something going on there.

Could it really be that neutrinos are moving faster than light?

Maybe. Theoretical physicists, particularly theoretical particle physicists, are nearly infinitely flexible. If another experiment finds the same result, I have no doubt that theorists will find a way to accommodate it.

It'd be deeply, deeply weird for lots of reasons, though, not least because the existence of superluminal particles that interact with ordinary matter (as neutrinos do, albeit weakly) opens the door to violations of causality—effects happening before the things that caused them, and that sort of thing. This wouldn't be a big loophole—the speed difference is tiny, and neutrinos interact extremely weakly—but it's the kind of philosophical problem that would really bother a lot of people.

So, if you have money to bet on it, bet that this result is wrong. But these guys aren't complete chumps, and if something is wrong with their experiment, it's something pretty subtle, because they've checked all the obvious problem areas carefully.

REFERENCES

The OPERA Collaboration. "Measurement of the Neutrino Velocity with the OPERA Detector in the CNGS Beam," http://arXiv.org/abs/1109.4897v1.

Siegel, Ethan. "This Extraordinary Claim Requires Extraordinary Evidence!" *Starts with a Bang* (blog), September 22, 2011. http://scienceblogs.com/startswithabang/2011/09/this_extraordinary_claim_requi.php.

CHAD ORZEL is an assistant professor in the Department of Physics and Astronomy at Union College in Schenectady, New York. He is the author of *How to Teach Physics to Your Dog* and *How to Teach Relativity to Your Dog*. He blogs at *Uncertain Principles*: http://scienceblogs.com/principles.

35

SUNRISE IN THE GARDEN OF DREAMS

PUFF THE MUTANT DRAGON

Gerhard Schrader had been feeling sick for a while, but he didn't realize how ill he was until he tried to drive home. There was something wrong with his eyes. He felt like he was going blind; he could hardly see the road, and it was becoming difficult and painful to breathe. It was a struggle for him to get home. When he finally reached the house, he peered into the mirror through his glasses, straining to see his own reflection. He discovered with surprise that the pupils of his eyes had shrunk to tiny pinpoints.

Chemist that he was, Schrader quickly guessed what had happened. Schrader worked at IG Farben, at that time, in 1936, the world's largest chemical company. He'd recently completed the initial synthesis of a new compound—a clear apple-scented liquid. Its sweet scent belied its real nature, however, for while he now knew the chemical's vapor must be quite toxic, he didn't yet realize how dangerous it was, or how close he'd come to an agonizing death.

After a couple of weeks in the hospital, Schrader returned to the lab just before Christmas, where he resumed work on his new compound—this time being just a little more careful. He soon realized his latest creation was incredibly toxic. Monkeys exposed to low vapor concentrations went into convulsions and died within minutes.

Schrader was heartily disappointed. He was trying to develop new pesticides; he wanted chemicals that killed insects but not mammals. Word of his unusual find got around to the German armed forces, or Wehrmacht, however, and unlike Schrader's managers at IG Farben, the Wehrmacht was deeply interested. As it turned out, in trying to find

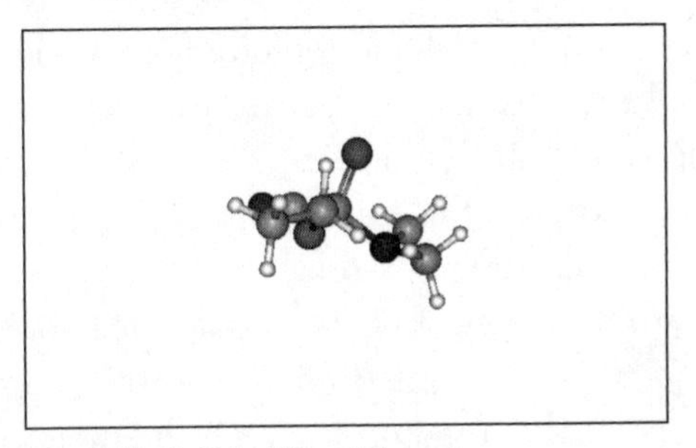

(Created by Puff the Mutant Dragon)

the perfect pesticide, Schrader had stumbled across history's most potent class of chemical warfare agents.

Luckily, neither tabun nor sarin (another similar compound synthesized by Schrader) saw use in the Second World War. Goebbels and a few top-level Nazis did try to persuade their Führer to use sarin, but Hitler declined, citing the risk of Allied retaliation. Despite the sinister nature of some of the compounds he made, however, Schrader's work ultimately proved beneficial, because his research gave rise to the modern world's most popular family of insecticides. Organophosphate pesticides (OPPs) are similar to tabun in many respects, and the way they kill insects is very similar to the way tabun nearly killed Schrader.

Certain OPPs have long been high on the list of controversial chemicals, and it's not difficult to see why. Nonetheless, these sophisticated poisons are invaluable to modern agriculture. That doesn't mean we can't eventually find a way to do without them.

The Master of All Poisoners

So how do OPPs work their lethal magic? Let's start by thinking about an insect as a biochemical machine we want to break. I know that sounds horrible—I like most insects, with the exception of cockroaches (especially when one turns up in a glass of orange juice I'm about to drink). But when it comes to pests that eat our crops, killing is exactly what we want to do. And just as you'd expect, it turns out some ways to put them out of action are more effective than others.

The bulk of the ammo we fire at pests is aimed at their nervous systems. That's hardly surprising; if you want to kill somebody, knocking out the nervous system is a fast and foolproof way to do it. Many of Nature's most lethal poisons act on the nervous system as well. Of the targets in the nervous system, however, the most popular by far is acetylcholinesterase (AChE). This, too, is unsurprising, and it has a lot to do with the way nervous systems work.

Every time you move your fingers or take a deep breath, your motor neurons convey the instructions from your central nervous system to your muscle fibers. At the junction between the motor neuron and the muscle fiber (the neuromuscular junction, or NMJ), the motor neuron floods the gap between the two cells (the synaptic cleft) with a neurotransmitter called acetylcholine.

On the other side of the cleft, acetylcholine binds to receptors on the muscle fiber, triggering a chain of events that causes the muscle fiber to contract. You don't want the acetylcholine to linger around, however, because that would keep the muscle in a contracted state. It's essential to get rid of the acetylcholine quickly—which is why you have the enzyme acetylcholinesterase.

Just thinking about Nature's catalytic tool kit is enough to turn chemists green with envy, and when you look at AChE, it's easy to see why. AChE boasts a turnover number of 1.4×10^4, meaning that when fully saturated a single AChE molecule can break down 14,000 molecules of acetylcholine per second. When acetylcholine enters the synaptic cleft, AChE immediately starts chopping it up, so as soon as the motor neuron stops dumping acetylcholine, the concentrations in the synaptic cleft drop quickly and the signal dies away.

This system generally works quite well, but from a strategic point of view, it's a weak point vulnerable to enemy attack. If you disable this

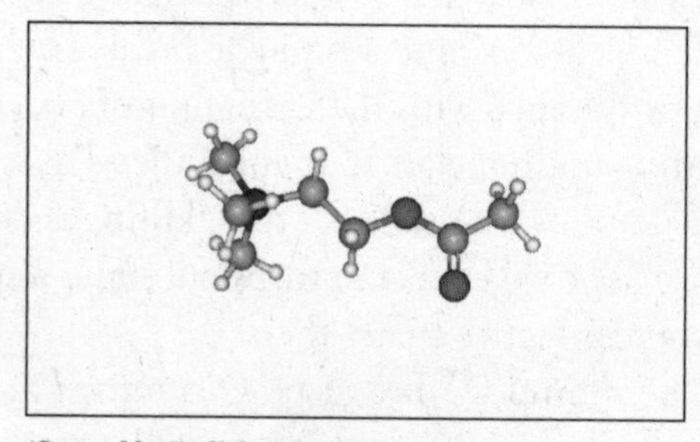

(Created by Puff the Mutant Dragon)

system, you deprive an animal of the ability to control its own muscles—you paralyze it. And Nature, the queen and master of all poisoners, has found ways to do precisely that. Botulinum toxin, or Botox, to take one famous example, paralyzes you by preventing your motor neurons from releasing acetylcholine. FAS2, a component of green mamba venom, relies on a different but no less vicious approach—it locks onto AChE itself.

We humans can hardly hope to surpass Nature's cruel cunning as a poisoner, but at least we can imitate her ways. Like FAS2, OPPs and nerve agents bind to AChE to knock it out of action. Let's zoom in and take a quick look to see how this works.

Below is a drawing of three amino acids ("residues") in the active site of the AChE enzyme from the Pacific electric ray. The translucent gray ribbons and loops are part of the cartoon drawing of the protein's secondary structure elements—alpha helices, beta sheets, and so forth. Everything else is hidden, except for the side chains of three particular amino acids. I'm not going to draw the full mechanism because some parts of it are still unclear, primarily the role of glutamate. We do know, however, that these three residues are critical to the enzyme's function. They serve as a kind of "catalytic triad," or trio, if you will. They're the cutting blades on the meat grinder, the teeth on the diamond-blade saw.

Acetylcholine fits right into the active site of AChE like a hand slipping into a pocket, bringing it into close proximity with the three residues

(Created by Puff the Mutant Dragon)

shown above. The interactions between acetylcholine and AChE subtly alter the shape of the enzyme and trigger a series of events that split acetylcholine into acetate and choline. Acetate is just acetic acid (vinegar) minus a hydrogen ion, while choline is a nutrient you get from your food.

Just like acetylcholine, nerve agents and OPPs also bind to the active site of AChE. Initially, the first few steps of the same reaction mechanism take place. But about midway through the reaction, something goes wrong.

Ordinarily, the oxygen on serine forms a bond with a carbon in acetylcholine, and this bond breaks during a later step in the reaction. When a nerve agent like sarin or an OPP is bound to the enzyme, however, a bond forms between the oxygen in serine and the phosphorus atom in the OPP, altering the normal sequence of events in a way that leaves the enzyme stuck. The end result looks like the image below:

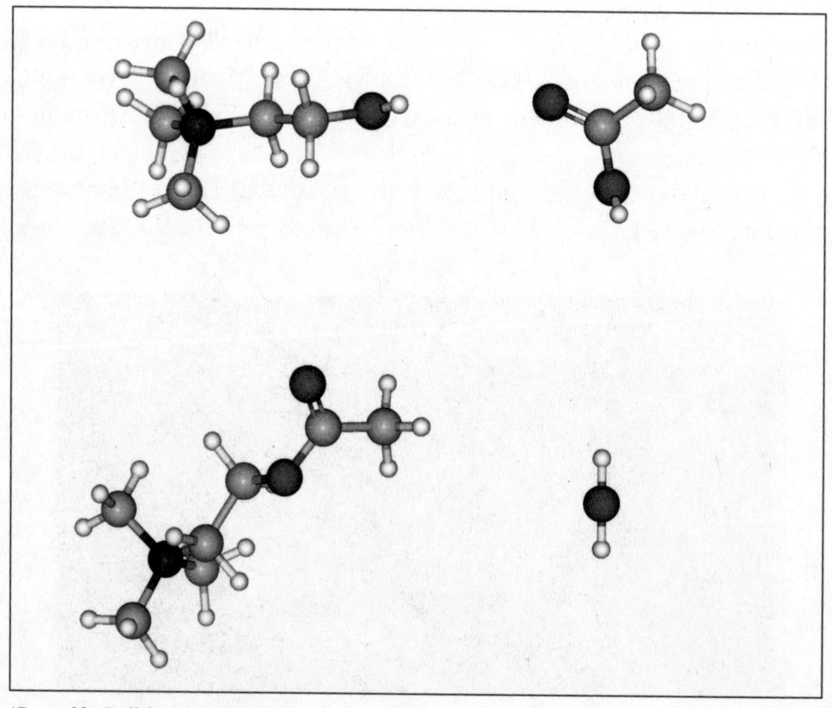

(Created by Puff the Mutant Dragon)

Notice the serine has a new group attached to it (a methyl phosphonate group). The result is the molecular equivalent of throwing a steel wrench into an intricate machine. With the methyl phosphonate stuck on the serine in the active site, the enzyme is disabled; it can no longer break up acetylcholine. The next time the motor neuron fires, acetylcholine will accumulate in the synaptic cleft with nothing to break it down. Even modest doses of OPPs condemn insects to paralysis and death.

Some pests, of course, have evolved varying levels of resistance to OPPs, usually through mutations that alter the structure of the AChE or alter other biochemical pathways that detoxify the OPP and render it harmless.

Another major class of pesticides called the carbamates also latches onto AChE, although the carbamates work in a different and less irreversible way. Nonetheless, they, too, are murderous to insects. Some carbamate pesticides like aldicarb are quite toxic to humans and other mammals, although aldicarb remains invaluable thanks to its potency against certain nematodes.

But wait just a minute here. If OPPs and carbamates are so toxic, how can we use them as pesticides? As it turns out, not all are equally toxic to all alike. Human and insect AChEs have different structures, so most of the OPPs/carbamates we use are more effective against insect AChEs than human AChEs. In other words, when compared with nerve agents like tabun, most of these chemicals are more toxic to insects and less so to us. That doesn't mean they are harmless, of course, and in fact some of them can be very deadly, but they are at least more dangerous to

(Created by Puff the Mutant Dragon)

insects than to humans. OPPs and carbamate pesticides are also thought to break down relatively quickly in the environment.

Chemical companies have occasionally erred in this regard. VX—the most deadly nerve agent of them all—was briefly marketed as an insecticide. The manufacturer soon discovered its startling toxicity and yanked the product off the market. Moreover, many of the OPPs and carbamates have long aroused bitter controversy, partly on account of evidence they may affect human/animal biochemistry through other mechanisms besides AChE inhibition. Many OPPs and carbamates can pose serious risks to farmworkers and potentially poison nontarget organisms like birds or fish. In 2009, for example, the U.S. Environmental Protection Agency banned carbofuran, a carbamate pesticide infamous for its lethality to birds.

Consumers often seem worried about pesticides in their food. It's worth noting that the U.S. Department of Agriculture tests produce samples as part of their Pesticide Data Program. Not only do a very small percentage of their samples actually test positive, but the levels of pesticides present are extremely low and well below the tolerance levels set by the EPA, which suggests this problem is far less significant than generally believed.

These kinds of controversies are difficult to resolve, however, not least because they reflect two different philosophies about how to regulate industrial chemicals. One of these is the precautionary principle, which holds that if there is any plausible risk associated with its use, a chemical should be treated as guilty until proven innocent in order to protect the public. Critics of the precautionary principle see this kind of approach as excessive. They prefer to consider these agents innocent until proven guilty.

A Poison Is a Poison

But wait! Do we have to use pesticides? Instead of arguing about which of them pose a health or environmental hazard, why can't we farm the "natural way" instead? That would be nice, if only a "natural way to farm" existed. Agriculture is inherently unnatural, a human invention that enables us to feed a population far in excess of what uncultivated land would ordinarily support. And when we plant vast tracts of land with crops bred to maximize yield, we are in essence making a giant Bug Magnet. We have to either control pests or accept significant losses. If we can find a different way to control pests, of course, then that's another story.

Critics of modern agriculture often cite "organic" farming as an alternative. The phrase "organic farming" secretly bothers me, because whenever somebody refers to "organic," the first thing that comes to mind is organic solvent. Nonetheless, misnomer though it may be, the name seems to have stuck, so we'll go with it. In any event, there are a lot of problems with the "organic alternative." For one, it's very dubious that humankind could live by organic farming alone. For another, organic farmers use pesticides, too. Sure, they're "natural" pesticides, but that doesn't always mean very much, because nicotine sulfate and rotenone aren't necessarily environmentally friendly agents either. At the end of the day, a poison is a poison is a poison, as Gertrude Stein might say.

Despite these drawbacks, there *is* a way we might be able to make organic farming viable in the long term. Unfortunately, it's the same approach the organic farming movement has already rejected. That's right: genetically modified organisms (GMOs).

How Green Was My Garden

There's a widespread misconception out there that natural=wholesome, good, healthy, while synthetic=dark, evil, designed by corporate gnomes to poison you. And there's some truth to this: recent history provides innumerable examples of cases where companies adopted practices later found to damage the environment. Think about tetraethyl lead or *Silent Spring* and you'll see why environmentalists are right to be on their guard. But at the end of the day, it's clearly a massive exaggeration to believe that everything large companies do is evil or that natural is greater than synthetic by default.

Unfortunately, this mind-set is still pervasive in the environmental and organic-farming communities, and it explains much of the hostility toward GMOs. To my mind, GMOs and organic farming could very well make an excellent partnership. I know there are a lot of reasons why this looks like a rocky marriage—Monsanto and organic farmers love each other about as much as monks loved Viking raiders. But we need to get past this stumbling block, because given further development, GMOs could ultimately prove to be the best thing that ever happened to organic farming.

I'm a little biased, it's true. I always find GMOs very exciting, because the potential is so vast. We're still in the early days; we're still just gearing up for action. There's a lot we have yet to learn about plant biology. But even the little we've accomplished so far gives you a hint of the promise to come. Nature's molecular tool kit is vast. If we can take greater advantage of that tool kit—use modern molecular biology to turn it to our own advantage—perhaps we can at last marry environmental stewardship to increased production. It's a beautiful dream, but today we might actually have the means to make it come true.

REFERENCES

Casida, John E. "Pest Toxicology: The Primary Mechanisms of Pesticide Action." *Chemical Research in Toxicology* 22, no. 4 (March 2009): 609–19. doi:10.1021/tx8004949.

Nelson, David L., and Michael M. Cox. *Lehninger Principles of Biochemistry*. 5th ed. New York: W. H. Freeman and Company, 2008.

Tucker, Jonathan. *War of Nerves: Chemical Warfare from World War I to Al-Qaeda*. New York: Pantheon Books, 2006.

USGS National Wildlife Health Center. "Organophosphorus and Carbamate Pesticides." Chap. 39 in *Field Manual of Wildlife Diseases: General Field Procedures and Diseases of Birds*. Madison, Wis: Biological Resources Division, USGS, 1999.

Zhou, Yanzi, Shenglong Wang, and Yingkai Zhong. "Catalytic Reaction Mechanism of Acetylcholinesterase Determined by Born-Oppenheimer ab initio QM/MM Molecular Dynamics Simulations." *Journal of Physical Chemistry* 114, no. 26 (July 2010): 8817–25. doi:10.1021/jp104258d.

PUFF THE MUTANT DRAGON is a Honalee-based freelance writer with a background in biology. In his spare time, he enjoys reading, blogging, and writing about science.

36

VOLTS AND *VESPA*: BUZZING ABOUT PHOTOELECTRIC WASPS

JOHN RENNIE

Wasps on the attack can seem like vicious little war machines: merciless flying Terminators, driven by cold insectoid rage, utterly unstoppable (except by a well-swung roll of newspaper). That comparison to robots may have ratcheted up a notch with a new report in the journal *Naturwissenschaften* that one species of hornet, *Vespa orientalis*, seems to be partly solar powered.

Like the worker castes of all wasps that live underground, those of Oriental hornets diligently expand their burrows by grabbing soil, flying a short distance from the nest to throw it away, then returning for more, perhaps hundreds of times a day. Entomologists have noticed, however, that Oriental hornets dig most feverishly at midday, when the sun is most intense, unlike other wasp species that concentrate their labors in the cooler early-morning hours. That unusual behavior prompted the late Jacob S. Ishay of Tel Aviv University to begin years of investigation into whether the hornets had some photoelectric traits that gave them an extra jolt of vitality in sunshine.

As far back as 1991, for example, Ishay found that shining light on the hornets—live, anesthetized, or even dead—could produce voltage differences of several hundred millivolts across their hard exoskeletons, which suggested that the cuticle material making up the exoskeletons was effectively an organic semiconductor converting light into electricity. Indeed, Ishay even found that shining ultraviolet light on an anesthetized hornet would wake it up faster, as though the light were recharging the insect. The general conjecture that emerged was that the distinctive bright yellow stripe across the Oriental hornet's otherwise brown abdomen might be functioning as some kind of solar cell, but the mechanism underlying the effect was unclear.

Carrying on Ishay's work, some of those same colleagues (and others) now think they have found confirming evidence for that hypothesis by looking closely at the structure of the cuticle with an atomic force microscope and other instruments. As they describe in the *Naturwissenschaften* paper, Ishay's former student Marian Plotkin of Tel Aviv University and her team found that the wasp's brown cuticle consists of an array of tiny layered grooves about 500 nanometers apart and 160 nanometers high.

In contrast, the yellow cuticle is made up of small interlocking protrusions, about 50 nanometers high, each bearing a "pinhole" depression. Both these sets of structures make the cuticle antireflective across much of the visible light spectrum: they absorb about 5 percent more light than a comparable flat surface would. Within the cuticle, stacked sheets of the hard polymer called chitin further boost the exoskeleton's ability to trap light.

The key element may be that the yellow stripe derives its color from the pigment xanthopterin. The researchers demonstrated that in an artificial solar cell, xanthopterin can work as the light-harvesting molecule, converting light into electrical energy with 0.0335 percent efficiency.

Plotkin and her colleagues suspect that within the yellow stripe, xanthopterin activated by the trapped sunlight separates electrical charges across layers of the cuticle; some other biochemical process could then tap this voltage gradient to make energy-rich molecules that the wasp's muscles or other organs could consume as needed. As Plotkin said to the BBC's Matt Walker, "We assume that some of the energy is transformed in a photo-biochemical process which aids the hornets with their energy demanding digging activity."

Of course, what that suggestion immediately brings to mind is the idea that the wasps are photosynthesizing. Photosynthesis, the ability to turn sunlight into biologically useful energetic compounds, is a trait that many of us still reflexively associate with green plants, algae, and cyanobacteria, not with animals.

That distinction has not strictly held for at least the past couple of years, since the 2008 revelation that the emerald green sea slug, *Elysia chlorotica*, draws energy from sunlight with the help of chloroplasts and genes picked up from the algae it eats. However, no biochemical mechanism yet identified in these hornets closes the loop and turns the energy in the cuticle voltage gradients into ATP, glucose, glycogen, or any other energetic molecule that could fuel cells.

But in 2009, Ishay, Plotkin, and their coworkers showed that, unexpectedly, a variety of important metabolic activities seem to center on the yellow abdominal stripes of the Oriental hornets rather than around the fat bodies that normally handle them in insects. (Think about what this means: if the same arrangement applied to humans, our skin would be doing the job of our livers.) Moreover, shining ultraviolet light on the yellow cuticle lowered the hornets' levels of several enzymes, including creatine kinase, alanine aminotransferase, and aspartate transaminase. So the yellow stripes do seem to be responsive to light in ways that affect the insects' metabolism, which might point the way to how the insects either produce energy molecules or use them more efficiently in strong sunlight.

All this work builds a strong circumstantial case that the Oriental hornets have indeed evolved an organic solar collector—perhaps not photosynthetic in the usual sense, but something similar. The answer Plotkin's team has deduced is so interesting, it makes me want to believe it. Ordinarily, I'd be loath to second-guess the experts' analyses and conclusions. And yet I must admit that certain points still perplex me about this discovery and leave me reluctant to embrace it.

First, the scientists' analysis doesn't quantify just how much energy a wasp could hope to gather by this mechanism. That 0.0335 percent conversion efficiency may be significant, but it sounds a bit paltry beside, say, the 0.1 to 2 percent efficiency that many plants show in converting sunlight to biomass. Of course, the wasps would be getting only an energy boost from the sun, not all their energy. For a wasp busily crawling through tunnels, dragging clods of soil to the surface, flying to dump them and returning, maybe every little bit of energy helps. But how much energy could it even theoretically recapture during its brief time aboveground compared to its overall energy expenditure? The antireflective and light-trapping properties of the wasp's cuticle do seem highly suggestive, but they might have evolved for other, unknown reasons, too.

Second, if exposure to light helps to charge up the hornets' metabolism, then why did Ishay's team find that ultraviolet light *lowered* the hornets' levels of those metabolically important enzymes below what was seen in the dark? I would think that diminishing the concentrations of creatine kinase, for example, would only interfere with the insects' ability to use or regenerate ATP in their cells.

Perhaps I'm simply being too cautious. Still, given the history of extraordinary findings that have sometimes dissolved under later scrutiny, it feels incumbent on all of us writing about unusual biology to be circumspect.

REFERENCES

Ishay, Jacob S. "Hornet Flight Is Generated by Solar Energy: UV Irradiation Counteracts Anaesthetic Effects." *Journal of Electron Microscopy (Tokyo)* 53, no. 6 (2004): 623–33. doi:10.1093/jmicro/dfh077.

Ishay, Jacob S., Ada H. Abes, Harry L. Chernobrov, Isaac (Zachy) Ishay, and Amir Ben-Shalom. "Electrical Properties of the Oriental Hornet (*Vespa orientalis*) Cuticle." *Comparative Biochemistry and Physiology Part A: Physiology* 100, no. 2 (1991): 233–71. doi:10.1016/0300-9629(91)90469-S.

Ishay, J.S., T. Benshalom-Shimony, A. Ben-Shalom, and N. Kristianpoller. "Photovoltaic Effects in the Oriental Hornet, *Vespa orientalis*." *Journal of Insect Physiology* 38, no. 1 (1992): 37–48. doi:10.1016/0022-1910(92)90020-E.

Plotkin, Marian, Idan Hod, Arie Zaban, Stuart A. Boden, Darren M. Bagnall, Dmitry Galushko, and David J. Bergman. "Solar Energy Harvesting in the Epicuticle of the Oriental Hornet (*Vespa orientalis*)." *Naturwissenschaften* 97 (2010): 1067–76. doi:10.1007/s00114-010-0728-1.

Plotkin, Marian, Stanislav Volynchik, Dganit Itzhaky, Monica Lis, David J. Bergman, and Jacob S. Ishay. "Some Liver Functions in the Oriental Hornet (*Vespa orientalis*) Are Performed in Its Cuticle: Exposure to UV Light Influences These Activities." *Comparative Biochemistry and Physiology Part A: Molecular and Integrative Physiology* 153, no. 2 (2009): 131–35. doi:10.1016/j.cbpa.2009.01.016.

JOHN RENNIE, a science writer, editor, and lecturer based in New York, is the former editor in chief of *Scientific American*. He writes *The Gleaming Retort* for PLoS Blogs and the Savvy Scientist column for SmartPlanet.com, and is an adjunct instructor in science journalism at New York University. Follow him on Twitter at @tvjrennie.

37

HOW TO STOP A HURRICANE

CASEY RENTZ

As another hurricane season passes, I'm surprised I didn't hear Bill Gates resound with his grand idea to dump tons of cold water in the path of such rotating monsters as Danielle, Karl, Lisa, and Tomas. Maybe Apple is plotting a more hip idea to unveil with the next iPhone release.

Despite a tech mogul's proclivity to bet on a solution, we know very little about how to stop a hurricane, though there are many quirky options on the table. I'll give you silver iodide, a supersonic jet, and raise you a nuclear warhead. You think I'm kidding.

Recent research says microscopic phytoplankton, of all things, might slow down the blustery beasts. These tiny photosynthetic plants, floating like flat, mile-wide bushes in the ocean, may drink up the sunshine that would normally warm waters and thereby feed hurricanes (cyclones), say scientists at the National Oceanic and Atmospheric Administration (NOAA) Geophysical Fluid Dynamics Laboratory in Princeton, New Jersey.

In the cyclone nursery, warm surface water evaporates quickly and, when it's reached the cold cloud cover above, condenses quickly. Droplets rain down on the open ocean. Winds blowing over the warm surface water encourage even more evaporation. Storms bluster forth. Earth's spinning gets the system turning, more heat is released by condensation, the energy of the thing builds, and *boom*—you've got yourself a rotating tropical cyclone, with a vertical engine churning out heat energy equal to exploding a ten-megaton nuclear bomb every twenty minutes. If phytoplankton could stop the warm air ascension and render a cyclone impotent, generations of researchers would love to see it in action.

Right now, only a computer model shows cyclone formation slowing down when it hits a patch of green. So how can that help us? Should we dump hundreds of gallons of phytoplankton in the 600-mile-wide path of a cyclone? The ecosystem effects could be devastating. And, with our current predictive technologies, there's no way to know exactly where this thing will go each minute. There are too many unknowns.

Most cyclone-stopping schemes never make it out of the theory stage for two reasons: logistics of putting such a scheme into action, plus unpredictability of the cyclone itself.

That's why many scientists were skeptical of Bill Gates's "dump a bunch of cold water on it" strategy. In 2009, the Gates-funded geoengineering company Intellectual Ventures proposed towing hundreds or thousands of buoys attached to gravity-powered "drains," which would sink warm surface water and stir up colder deep water, into the path of a potential cyclone. But cyclones pop up so quickly—one hour it's just a spot of low pressure, the next hour it's halfway to cyclone status—that airplanes or barges doing the towing would have to be on higher alert than Canadian snowplows in January. It would be extremely difficult to get dozens of buoy-laden helicopters or ships out over the ocean and into a precise location within twelve hours. (Air traffic control would be a nightmare.) And even if you knew where the cyclone was going (not exactly possible with current technology), you would have to cover an estimated sixty square miles just in front of it, says Bill Gates's team. If you weren't sure which path it would take, you would have to cover at least three sides, tripling the cost and the coordination efforts. And even then, you're just slowing it down a bit.

Though Gates's idea sounds too cumbersome and costly to be worth the effort, other solutions sound like they're ripped right out of a comic book.

Last year, fluid dynamics engineer Arkady Leonov of the University of Akron suggested sending supersonic jets careening into the eye of a cyclone and having them circle around and around, against the flow of its winds. According to Leonov, the winds and sonic booms from the jets might slow the storm, and flying near the water could cut off the circulation of warm air from below. NOAA scientists are skeptical. An equally possible scenario is that the jets get torn apart by the cyclonic wind or run into each other; or if they do survive, the wind they create might make no difference at all.

NASA jets flew into Hurricane Earl in 2010, just to observe the cyclone in action. But NOAA has been sending aircraft into cyclones for about thirty years now. In the 1960s and 1970s, NOAA ran "Project Stormfury," dedicated to disrupting the eye of a cyclone—like targeting the engine of a car—with a substance called silver iodide.

Inside a cyclone, bands of rain layer from the center (the eye) outward and house supercooled water. Silver iodide forms ice nuclei with supercooled water, causing it to freeze. Heat is released when molecules fuse in the freezing process, and the rain band grows, collapsing the cyclone's inner eyewall. Scientists actually tested the silver iodide method, dropping it into Hurricane Debby in 1969. The cyclone was weakened, but only temporarily. This all happened before anyone knew that cyclones go through natural stages of weakening and strengthening, as outer storms replace the inner ones close to the eye. Lesson learned.

So, if we can't freeze a hurricane out, can we dry it up instead? In 2001, Dyn-O-Mat Company patented a water-absorbing substance called Dyn-O-Gel, similar to the stuff used in the absorbent strip of a baby's diaper. The company suddenly got all pie-in-the-sky about their invention, claiming Dyn-O-Gel could suck the moisture right out of a moving cyclone. "This powder will give you perfect weather every day," said the company spokesperson in *Woman's World* magazine.

The company actually tested it on live clouds and storms in the Caribbean. When they contact water, the small particles are able to absorb water up to several thousand times their own weight and create a heavy gel. The clouds and thunderstorms tested actually did seem to disappear. But the study methodology was somewhat questionable: there was no "control cloud" to see if they would have dissipated anyway. And Dyn-O-Gel has the same cost-logistical problem as Bill Gates's cooling buoys. Dyn-O-Mat's proposed "2000-to-1 Dyn-O-Gel to water" ratio means that a typical 1,500-square-mile cyclone would call for more than 30,000 tons of goop. That much goop would call for 300 heavy-load aircraft, at 100 tons each, dumping their load every two hours. Ridiculous.

Okay. We can't slow its turning winds, we can't lower the surface water temperature, we can't wipe out its eye, and we can't dry it out. How about we prevent evaporation of the tepid water in the first place? This solution was suggested back in 1966 and again in 2002—just pour some oil or other surfactant around the cyclone, and the water would be trapped below the slick. But, alas, most substances separate into pools, and evaporation persists in the spaces in between. Has anyone tried to just cover up the water with plastic wrap?

Now we're getting desperate. If we really want to control the destructive powers of a cyclone once and for all, why don't we just nuke the bastard? Every year, it gets mentioned somewhere. Desperation is not pretty, and expensive at $2 million–$10 million per bomb.

In fact, the projected cost of any of these heroic efforts—water buoys, tons of silver iodide, gobs of Dyn-O-Gel, farms of phytoplankton—is through the roof. And none of them promise to stop a cyclone completely, just slow it down. It leaves me wondering: is it cheaper to just rebuild a coastline than protect it from destruction? Maybe. But, as Katrina proved, our nation can't even commit to rebuilding after a major disaster.

The one thing that might help move these big, heroic, imaginative ideas into the realm of the possible is to develop better predictive technologies. Having a "cyclone compass" might make Gates's cooling scheme more practical—we could think about towing in 200 instead of 600 cold-water buoys if we knew exactly where to drop them. And the less oil or phytoplankton used, the lower the costs. In the meantime, predictive meteorology can save lives by providing more advance notice for an impending coastline evacuation.

So maybe it's systems for prediction rather than prevention that lie in our immediate future. We need to know where cyclones are going before we can control them, right? It's all part of Microsoft's plan to manipulate weather on Earth so the AT&T cell signal for the iPhone gets worse and worse. But for now, we have to admit we're dealing with a force of nature that, with all our technological advances, we cannot control. Yet.

CASEY RENTZ is a freelance science writer living and working in Los Angeles. After years of researching tiny phytoplankton and concocting art projects on the side, she enjoys the fruitful marriage of science and art that science writing begets.

38

SHAKES ON A PLANE: CAN TURBULENCE KILL YOU?

ALEX RESHANOV

Airplanes are scary. This indisputable fact originally came to my attention sometime during college, when my relationship with flying matured from irritated ambivalence to full-blown phobic terror. There is something profoundly disturbing about being suspended thousands of feet in the air in a metal tube piloted by someone you've never even met.

When I explain this to people who aren't bothered by air travel, they foolishly try to persuade me with statistics about how flying is safer than driving, blah, blah, blah . . . as though irrational fears could be soothed by something as banal as data. Really, the only way to take my mind off my impending death is to pretend I'm not in an airplane at all.

Distractions like food and music and in-flight magazine crossword puzzles go a long way toward accomplishing this, but all their hard work is undone the instant the plane encounters even a little turbulence. Turbulence has a way of snapping you back into the present moment, its every lurch and bump an unwelcome reminder that you're hurtling through the stratosphere at over 500 miles per hour.

My spring vacation involved a total of seven flights and twenty-five hours of time in the scary skies. Mercifully, every flight was smooth and trouble free, except for the very last one—a short jaunt from Dallas to Austin.

I could tell the final phase of the trip was going to be less agreeable than the previous portions. Thick clouds hung over the airport while we waited to board the plane. As soon as we were in the air, the pilots announced that there would be no beverage service due to some rough weather ahead. Throughout the forty-minute flight, we received various ominous announcements from the cockpit as the plane shuddered in increasingly malevolent winds: "We're going to fly low today and try to stay below this storm," and "Okay, so still pretty bumpy even at this altitude, but we should hopefully be there soon."

After we were safely back on the ground, I apologized to my boyfriend for all my in-flight whimpering, calling upon the mantra I'd heard dozens of times: "I know it's just turbulence and it can't hurt the plane, but it really feels like you're about to crash." Instead of laughing at me for being such a sissy, he launched into a lengthy discussion of how severe turbulence *can* cause plane crashes, and how the turbulence we'd just experienced had been pretty rough, and something about "wind shear" being capable of tearing planes in half. I was really glad he hadn't shared any of this with me during the flight.

Say It Isn't So

Since anecdotal boyfriend babble is not always accepted as a reliable source, I did some research on whether turbulence can cause aviation disasters. The answer turns out to be a heavily caveated yes. Turbulence can lead to plane crashes, but it is exceedingly rare. By some estimates, turbulence takes down about one plane per decade.

This chaotic air movement—and its effects on the movement of the aircraft—is classified in degrees of light, moderate, severe, and *extreme*. There's also something called "chop," which is a more rhythmic bumpity-bumpity effect that comes in light and moderate flavors. Passenger perception of turbulence tends to be direr than that of experienced crew, so if you think you've been on the worst flight of your life, it's likely you encountered only moderate turbulence.

The good news is that airplanes are designed to withstand extreme turbulence (as well as lightning). The bad news is that, like all machines, airplanes age. Wear and tear that is no problem under normal circumstances can make aircraft less resilient to ridiculous levels of turbulence.

Additionally, flying a plane that is being pummeled by rogue air masses isn't the easiest thing in the world. Planes can be pretty much out of control during these episodes, and, while they're temporary, the pressure is really on the pilot to react (but not overreact) in a way that keeps the aircraft from flying into the side of a mountain. If you think this all sounds melodramatic, consider a 1966 incident in which a BOAC (now British Airways) Boeing 707, flying near Mount Fuji, broke up in midair and crashed amidst harsh winds.

Your Wake, My Funeral

As frightening as that is, you should probably be more concerned about something called "wake turbulence." Unlike turbulence created by naturally occurring differences in airflow (bad weather, pressure variations near mountains, jet streams), wake turbulence is caused by other airplanes, sort of like the wake created by a boat, except with air and much scarier. The worst part of wake turbulence is the creation of "wingtip vortices," tornadoes of bumpy air generated by a plane's wings that can take several minutes to dissipate.

As with boat wake, this form of turbulence is not a problem for the planes creating it, but rather for the planes near it. Airports enforce strict limits regarding how much time must pass in between takeoffs and landings to prevent one plane from getting caught in another's wake. However, airports are increasingly crowded places with limited runway resources. Since they are traveling at slower speeds and pitched at awkward angles, planes that are taking off or landing are more vulnerable to turbulence. And, unfortunately, takeoff and landing are exactly the times when planes fly closest to one another.

As with other forms of turbulence, wake turbulence is more of a threat to small planes, especially when caused by the significant wakes of large commercial jets. Probably the largest aircraft to crash as a result of the phenomenon was a McDonnell Douglas DC-9 that got caught in the wake of a Lockheed L-1011 in 1972, prior to the implementation of the above-mentioned spacing regulations. Wake turbulence allegedly also contributed to the 2001 demise of American Airlines Flight 587, which plunged into a neighborhood in the New York City borough of Queens shortly after takeoff, though this crash is officially attributed to pilot error in response to the wake.

Reality Check

And now let me stress how very, very rare these occurrences are. Finding examples with which to freak you out was no easy task. Really, the biggest of your worries in the realm of bumpy air is encountering clear air turbulence (CAT) when you don't have your seat belt fastened. This is the surprise turbulence that occurs on a sunny day in seemingly smooth skies. The major impact of turbulence is not to cause plane crashes, but to cause bodily injury with all that jerking around. Every year, dozens of unbelted passengers are seriously injured (and occasionally even killed) by being thrown around the cabins of twitchy planes.

Your pilot isn't just being an overbearing nag when he (or she) asks you to keep your seat belt fastened when you aren't walking around the cabin. CAT can come out of nowhere and knock the #%$ out of a plane. So buckle up, people.

ALEX RESHANOV is an Austin-based writer and the creator of *Blogus scientificus,* where diverse science topics mingle awkwardly with dubious puns. She is also a regular contributor to *EarthSky.* You can read her latest posts at http://blogusscientificus.blogspot.com.

39

HOW ADDICTION FEELS: THE HONEST TRUTH

CASSIE RODENBERG

It's hard to confess that I have a problem with sleep, but I do. I'm a narcoleptic. Some days I feel like an alcoholic—unable to think of anything else but the craving. Sleep and I have a tumultuous relationship. I'm constantly exhausted and wanting more, more, and will never turn down a nap. On the other side of the coin, or pillow, I despise it.

The current medical answer for me? Downing a combination of stimulants and antidepressants, carefully enabling me to be awake and alert while only minimally zombie-like. The short answer is there is no answer.

Does the need-hate relationship sound familiar? The same is true for addiction. (To note: narcolepsy is not an addiction, but I'm using the similarities to describe the feelings of addiction as best I know how.)

Until now, I've made it a policy to tell only the closest people in my life that I have narcolepsy, though this would likely benefit all of my acquaintances (i.e., if my eyes glaze over, it's not you). I do this mostly to avoid the bug-eyed looks and discomfort, and the question "Do you fall asleep standing up?" Answer: Yes, I have. No, I do not do so often.

Hiding It: Shame and the Emotional Cycle

Most addicts harbor an urge to hide what feels like a terrible secret, and so do I, though it's just biology.

Worse than hidden mental battles are those that scoff. Various relatives have fallen into this category and have tried to shame me awake over the years: "If you loved me, you'd want to spend time with me while you're visiting; you wouldn't want to sleep."

This ignites screams and fire in my head. Emotional blackmail doesn't work on a neurological disorder any more than it works to tell someone with a heart condition that, if he loved you, he could walk up a flight of stairs without resting. (To be fair to families and friends everywhere, brain trouble isn't as simple to understand as heart trouble. It takes time.)

The irony: when relatives play the emotional hand to encourage me to stay awake, my stress level spikes, which triggers my need to sleep. Addicts feel an overwhelming need to use in stressful situations, and then they feel irrevocably guilty, since they really don't want to use the stuff in the first place. Guess what? I would love to be awake. Addicts would dearly love to lose their compulsion to use.

Daily Life

Like an addict, I spend my days trying not to think about the thing I crave, trying to avoid situations that set me off. Situations like being near a bed, or the dreaded 1:00–3:00 p.m. block during the day, or just thinking about sleep, might be unique triggers for narcoleptics, but other compulsions are not. For instance, emotional situations, either good or bad (family turmoil, a great new job), send addicts spiraling. At such emotional cues, for me, it's as if the universe's force momentarily manifests to drag my eyelids closed.

As you might guess, the constant urge to sleep (or to use) leads to interesting psychological encounters and relationship problems over the years, and at times I've questioned myself and my sanity. Whether it's sleep or cocaine, the feeling is that there's something badly wrong with you when you go on a drug binge around the holidays or fall into a coma-like slumber after a fight.

My psychiatric conversations can be summarized neatly like this:

PSYCHIATRIST: A lot of people who are depressed feel the urge to sleep. Do you feel depressed, or sad?

ME: No, I'm not depressed. I just need to sleep.

PSYCHIATRIST: That might mean you're depressed. It's perfectly okay to feel depressed.

ME: I know it's okay. But I don't feel depressed. I'm pretty happy in life.
PSYCHIATRIST: I think you're denying the fact that you're depressed.

"Circular," "tiresome," and "discouraging" are mild terms to award the frustration I've felt. Unfortunately, many a therapist isn't trained to counsel and treat addiction or sleep disorders. Diagnoses are hard and can take years, especially when there isn't a cure-all solution.

Really, there could be any number of things wrong. Many addicts begin therapy for what are called co-occurring disorders, like addiction and depression. Addiction sometimes bleeds into other problems, or trades one form for another, such as opiate abuse for an eating disorder. Sometimes medication makes you feel more unhinged than before, or like another person entirely. You start to lose sight of yourself and who you are. Is it "you" when you use illicit drugs all the time? No. Is it "you" when you feel like a robot on treatment medication? It takes time, patience and personal exploration to find the right balance. There's no quick fix.

Work Balances

Addiction can take a toll on your career, and actually frame the shape of it, in less-than-obvious ways. It's more than missing work or punching the time clock while under substances' influence: it can be a problem with your industry and your work itself. I've known pharmacologists who handle pills they abuse all day, or restaurant servers who ferry customers alcohol. There are ways to get around these things (having someone else physically handle the prescription drugs, asking someone else to serve spirits), but it's something that rests perpetually on the brain.

For a long time, I was intent on going to medical school; finally, and after a long personal game of tug-of-war, I chose not to go. The major reason? I didn't think I could stay awake during the infamous round-the-clock shifts of medical school. I was scared and didn't want to ask for special treatment, knowing the internal politics of hospitals.

Long term, I didn't think I could handle the lengths of shifts or the uneven schedules of shifts without rest. I still think about trying it, going back to study medicine. In my heart, I know it would never work, at least not in the way I would want it to. Addiction can actually change lives and careers; this disorder has changed mine.

Day-to-Day Management Tactics

It isn't all bad, and I live a happy, full life. But it should be said how aware we have to be. That's the word: aware. I lump myself in with addicts here. What do I have but a sleep addiction? (Actually, addiction and narcolepsy are related genetic disorders.) I have to plan my days through a different veil than most people, and have an exit plan in place. I'm going to a party? I'll get there early so that I can leave if I feel tired, without seeming rude.

Some days and events coalesce into situation roulette. Sometimes I might be able to stay up until 2:00 a.m. without any more of a problem than the next guy. Sometimes an addict might be able to walk near the beer aisle safely. The cravings' timing varies and is malleable, which makes the disorder harder to understand. At times, after five minutes of being the living equivalent of the walking dead, I can fight through it.

Addicts all have trigger cues, or settings where they're more likely to use. Sometimes alcoholics need to leave a party when champagne appears. I've had panic attacks when I felt too far away from a place to sleep—once at a crowded, noisy rock concert thousands of miles from home.

What It All Means

Addiction is confusing, inconsistent, and inconvenient, just as much so to addicts as to family and friends.

A friend once asked what reaction I'd like from people, or how he could help. Giving love and understanding helps. Trying to understand what I'm going through helps, because sometimes I don't know how to express how it feels. Like describing an emotion, sometimes it's hard to form in words, especially if your audience can't fathom feeling the same way. The point is to try.

I've nearly ruined relationships by my inability to explain my disorder, by hiding it, and by just not knowing. I didn't know then what I know now. Being explanatory, honest, and open to questions converges as the pinnacle of success for most things—this is no different.

I think I can speak for most addicts in a note to family and friends:

We do love you, and we're sorry we put you through this hard journey of ours. We don't mean to, and we're trying to figure out how to work on it. The trouble is, no one really knows how. But don't write us off yet. For everything over the years, thank you.

CASSIE RODENBERG is a chemist turned writer and digital producer who works and lives in New York City. More than 10 percent of the population, and much more than 10 percent of the people in her life, struggle with addiction. She writes to make clear the science and to crack cultural taboos. Addiction and mental illness is *The White Noise* behind many lives—simply what Is. Find her at http://blogs.scientificamerican.com/white-noise/.

40

TEN MILLION FEET UPON THE STAIR

CHRIS ROWAN

During my time in Edinburgh, I lived in an apartment in a nice old tenement building: several floors of individual flats, all connected by an internal communal staircase. The building is at least a century old, and because this was back when things were built to last, each stair is a slab of gritty sandstone. But nowadays, each of those slabs is looking a little worn.

This wear is the cumulative result of a century of people walking up and down from their flats. As they left for and returned from work, as they nipped out to the shops or ventured out for an evening in one of Edinburgh's many pubs, many times a day the feet of the people who lived here would fall upon each stair.

The force applied by each footfall may not be great, even for those who had overindulged in deep-fried Mars bars.* But as every geologist knows, even a small force, repeated over a large enough stretch of time, can add up to some very large effects indeed. And a brief (and very rough) calculation shows that a century's worth of footsteps is quite a lot:

- Assume 30 people living in the building at any one time—a very conservative assumption for 16 two-bedroom flats, especially since in the early twentieth century this tenement was probably a lot more crowded.
- Assume each person travels up and down the staircase 4 times a day on average, thus stepping on each step 8 times a

*The deep-fried Mars bar is the semimythical apotheosis of the alleged Scottish propensity to deep-fat-fry everything they eat.

(Photograph by Chris Rowan, 2010)

day. This comes to a total of 240 footsteps per step per day—factoring in visitors and postmen, let's say 260.

- Multiplying up, this comes to 1,800 footsteps per step per week, around 7,000 a month, or almost 100,000 a year.
- Therefore, if the building was just over a century old (which is a reasonable estimate for the area of Edinburgh I lived in), each step has been struck by at least 10 million feet since it was laid down.

When I climbed this staircase to my flat, I would often look at the uneven steps and ponder this visible sign that my home had a history, that I was merely the latest in a long line of people who had lived there, some before I was, or even my grandparents were, born. Indeed, in a city like Edinburgh, feeling a bit parvenu is a common occurrence.

But, perhaps unsurprisingly given my profession, I also enjoyed the daily reminder that humans are a geological force; not individually, perhaps, but through sheer weight of numbers. If each of those 10 million feet were attached to a person, and those 10 million people were all sent up this staircase in single file, it would take less than 8 months for them to collectively erode away a centimeter of sandstone.

And that's when they're acting in series—passing up one by one. What if they could act in parallel; if, somehow, they could all step on the same stair at the same moment? Each of those 10 million light footsteps would add up to a sledgehammer blow that would, in an instant, turn that centimeter of rock to dust.

And then, consider that 10 million people is but a small fraction of the almost 7 billion people currently in the world. If you could somehow harness the feet of all of those people at once, then you could grind 7 meters of rock away in the blink of an eye. A few more repetitions and you'd have an impressive ravine. Keep going for a few hours, and you could produce a new canyon.

All of this might seem like a rather implausible thought experiment, and it is certainly rather impractical to have the entire population of the world planting their footprints on a single stair. But when it comes to our carbon footprints, the entire planet is the stair.

Our individual contributions—the energy we expend, the waste we produce—may seem insignificant, hardly something that is going to affect the planet. But when you multiply it by 7 billion, the small environmental impact of any one person becomes a very weighty footstep indeed; and after more than a century of increasingly heavy footfalls, it's not surprising that Earth is as worn down as my old staircase.

In his day job as a geologist, CHRIS ROWAN studies the movement of plates and the deformation of continents, and still hopes to succeed in his career without having to grow a beard, which would not become him at all. He writes about earthquakes, plate tectonics, and other geological topics at the blog *Highly Allochthonous* (http://all-geo.org/highlyallochthonous).

41

WOOF! LIVING BOLDLY AS A "FREE-RANGE ASPERGIAN"

STEVE SILBERMAN

John Elder Robison would stand out in a crowd even if he didn't have Asperger syndrome. A gruff, powerfully built, tirelessly curious, blue-eyed bear of a man, he hurtles down a San Diego sidewalk toward a promising Mexican restaurant like an unstoppable force of nature. "What's keepin' you stragglers?" he calls back to the shorter-legged ambulators dawdling in his wake.

As they catch up, Robison utters his all-purpose sound of approval—"*Woof!*"—which he utters often, the benediction of a man in his middle years who is finally at peace with himself after a difficult coming-of-age. For the acclaimed author of the 2007 *New York Times* bestseller *Look Me in the Eye*, a diagnosis of autism spectrum disorder in midlife was liberating, giving a name to the nagging feeling that he was somehow different from nearly everyone around him.

Like many people with Asperger syndrome, Robison considers himself, as he puts it, "an aficionado of machines." He built his first DIY creations with an Erector Set. The elegant machines that sustain his livelihood now have names like Rolls-Royce, Mercedes, Land Rover, and Bentley. His high-end auto repair shop in Springfield, Massachusetts, is one of the most successful independent body shops in New England.

As a precocious gearhead in the late 1970s, the bearded, bespectacled Robison was the guy who pimped out Ace Frehley's guitars to spit flames, spew smoke, and erupt in cascades of light at KISS shows, sending the band's legions of admirers into spasms of pyrotechnic ecstasy. A veritable rock-and-roll orgy was under way all around him, but Robison (who had a girlfriend at home) was too terrified to even start up a conversation with a woman he didn't know. After his time on the road, he took a job at Milton Bradley developing talking toys and games. Then came a stretch of unfulfilling years supervising sales of laser power supplies and fire alarms. Finally, in 1986, he opened his auto repair shop.

In *Look Me in the Eye,* Robison described why he felt more comfortable around machines than people: "No matter how big the machine, I am in charge. Machines don't talk back. They are predictable. They don't trick me, and they're never mean. I have a lot of trouble reading other people. I am not very good at looking at people and knowing whether they like me, or they're mad, or they're just waiting for me to say something. I don't have problems like that with machines."

One day, a longtime friend who was also a therapist handed him a copy of Tony Attwood's *Asperger's Syndrome* and said, "This book describes you exactly. You could be the poster boy for this condition." Robison was skeptical for only the few moments it took him to glance through the pages and realize that his friend was right. Robison asked him if there was a cure. "It's not a disease. It doesn't need curing," his friend replied. "It's just how you are."

Discovering that other people struggled with the same sensory overload, inability to read faces, anxiety, and single-minded obsession with topics of interest helped catalyze Robison's transformation into a writer. Bestsellers run in Robison's family; his younger brother, Augusten Burroughs, published his own account of their troubled upbringing called *Running with Scissors,* which was turned into a movie. The Aspergian frankness and guilelessness of *Look Me in the Eye* gave the book its distinctively bracing candor, much as Holden Caulfield's X-ray vision for the "phonies" of Pencey Prep did for *The Catcher in the Rye.* Robison next wrote *Be Different,* a book of anecdotal wisdom for young people on the spectrum, with the pragmatic message to revel in your autistic differences but also to try to fit in as much as you can.

It's a book about hope—but not the kind of hope touted at conferences with the promise of "recovering" your child from autism. Instead, Robison counsels self-acceptance with strategies for adapting to a world designed for nonautistic people. "The brain differences that make us Aspergian never go away," he writes, "but we can learn two important things: how to play to our strengths and what to do to fit in with society. Both those skills will lead to a vastly improved quality of life."

The neurologist Oliver Sacks famously compared the life of the autistic author and livestock-industry consultant Temple Grandin to that of an anthropologist on Mars. In some ways, Robison's books are like travel guides to the strange customs of Earth for those who feel like they were born on the wrong planet. "I've learned to say 'please' and 'thank you' fairly often," he says in *Be Different*. "That's a simple rule that delivers good results."

I spoke with Robison about the aspects of life on the spectrum that he still finds challenging, what it's like to be one of the few autistics in an advisory role to a fund-raising organization like Autism Speaks, and the folly of branding people either "high-functioning" or "low-functioning."

STEVE SILBERMAN: *What inspired you to write* Be Different?

JOHN ELDER ROBISON: When I wrote *Look Me in the Eye*, I never intended it to be an all-inclusive guide to autism—it was just the story of my life. But so many people have come up to me asking me to explain how I became successful. They tell me, "You said that you were going to teach yourself to fit in and you did. I want to know how to do it myself."

Then other people say things like, "I don't understand how you could claim to be a person with autism and yet be at these loud rock-and-roll shows with flashing lights. My son can't stand anything like that." They want to understand how people with autism can be so different from one another and yet the same, and if I have some secret that will help their kid.

I don't have all the answers. But the vast majority of people who read my stuff have a personal stake in autism—whether it's them, their husband, their boyfriend, their child, or people they work with at school. That made me realize that I have a duty. When somebody asks me "Why?" I shouldn't just shake my head and say "I don't know." *Be Different* was the result of a journey of unraveling why I do the things I do and why I feel certain ways.

I thought very carefully about the things that have made me successful, and also the things that sometimes keep me from being successful. I looked at the DSM criteria for autism and Asperger's and considered how each trait affected me. Some of the traits are just purely crippling things; there's no good side to them for me. But other traits, like social disability . . . if I didn't have friends because I was isolated by disability, I also had free time to study what I was interested in, which was technology.

I'm not very flexible—all I do is think about electronics and computers—which people *think* is a disability. But when you combine that with my autistic fixation on things, you've got a kid who has a really powerful ability to learn about what he's interested in. You can go a long way with that. Some traits that cripple me in one way have been a gift to me in another way.

SILBERMAN: *What aspects of autism do you still find really hard to deal with?*

ROBISON: The social disability is the most challenging. As a middle-aged adult, the biggest danger to me is falling into a serious depression. That happens to me more or less exclusively as a result of my social disability. That's the hardest thing for me. I can only go so far in remediating that. You know?

If you take me and a mom with a newborn baby and sit us in front of a computer and flash faces in front of us so we have to push buttons to say what the faces mean, I don't have any idea. But the mom can get those things right most every time. That means that when you put the two of us into a strange social situation, the mom can just read the room. I can't. But I've taught myself adaptive behavior. I look at people slowly and think, "What's he looking at? What's he doing and what does it mean?"

By now, I can make pretty good judgments about what people are feeling toward me. But it takes a lot of energy. Ultimately, I can only make a logical determination. But the mom can look at somebody who's angry and feel his anger and pain. I look at the person who's angry and it's a kind of dispassionate, detached process. I think, "Hm, he's angry—I wonder why?" I don't *feel* it. So that causes me to not be able to respond appropriately in some situations. Frankly, it still causes a great deal of stress for me.

SILBERMAN: *Are there any ways that society could be reformed to make it a more comfortable and supportive place for autistic adults?*

ROBISON: I don't think that's a realistic question. We represent 1 percent of the population. Asking what 99 percent of the world should do to make it a better place for that 1-percent member—that's verging on science fiction and fantasy.

If you're a guy with severe autistic disability and you can't talk, you cry out for compassion by your very existence. It's obvious when people look at you and listen to you. If you're a person in a wheelchair, nobody can reasonably argue that you should just get your ass across the street.

But when you're a person like me and your disability is principally with social functioning, and at the same time you have good language skills, people are just going to dismiss you as a jerk if you don't learn to fit in. That's the hard truth.

SILBERMAN: *You weren't diagnosed until you were forty. What did learning you're on the spectrum in middle age do for you?*

ROBISON: It showed me, for the first time in my life, the underlying cause for my exclusion from society. That was a tremendously empowering and liberating thing. Before that, how could I ever know that everyone else isn't just the same as me? People would say "*Look* at me, John," and I believed I was fully complying with that request. Of course, in the opinion of other people, I wasn't complying at all. But I had no way to know that. So that's an example of how diagnostic knowledge can be tremendously empowering. You really have no potential to have a good life if there's some fundamental difference between you and everyone else and you don't understand what it is. There's no way you're going to integrate yourself with everyone else in ignorance.

SILBERMAN: *In the next edition of the bible of psychiatry, the DSM-5, scheduled to be published in 2013, the diagnosis of Asperger syndrome is probably going to go away. Instead, the umbrella of autism spectrum disorder will be broadened to include many folks like you—though precisely how many is still in dispute. What's the virtue of emphasizing the continuum between people with Asperger and people with so-called classic autism?*

ROBISON: The biggest virtue is that we recognize that people can occupy different positions on the continuum at different points in their lives. One of the things that troubles people about the use of labels like "low-functioning" and "high-functioning" is that people will call a five-year-old kid who can't talk "low-functioning," yet a kid who has language skills, like me, but doesn't have any friends, is described as "high-functioning."

First of all, of those two children, the so-called high-functioning kid is the one who is at material risk for suicide by the time he's sixteen. Most people would not call a dead kid highly functional.

Then, when we look at the so-called low-functioning kid—a large number of those kids, given the benefit of intensive speech and language therapy, acquire language skills that are indistinguishable from so-called ordinary children by the time they're eighteen. So having the low-functioning label applied to that kid just serves to demean him, hold him back, and minimize the expectations that others have for him. At the same time, I recognize that there are some kids who have severe impairment when they're five and have severe impairment when they're eighteen, and it doesn't change.

If we had to distinguish, how could we pick out the kids who will not grow out of their impairment? Studies show intelligence is the most accurate predictor of one's ability to emerge from disability. But we cannot accurately measure intelligence in a child who does not have a reasonably age-appropriate power of speech, because without that, a kid can't be expected to comprehend the questions on the IQ test.

SILBERMAN: *Right. And standard intelligence tests are biased toward neurotypical modes of perception and interaction.*

ROBISON: Absolutely. If you can't talk, you can't get anything other than a seriously subnormal score. But some of those kids don't have low IQ, they just have a severe language challenge, and we don't presently know how to distinguish that among four-year-olds. So that means that we do not really have the ability to make any kind of prediction about which children will emerge from disability and to what degree. A lot of it also depends on their motivation.

SILBERMAN: *What was the most surprising thing about meeting other autistic people once your first book was published?*

ROBISON: One thing that was really surprising was the way in which a lot of those people come to my events and sit right in the front of the audience and make noises and gestures that I can't understand. Yet those people are obviously paying rapt attention when I'm speaking, and it's clear to me that they're getting something of value from coming to my talks, because they show up again when I come back to town. Those people supposedly have little or no language, but they're getting something or they wouldn't have a reason to want to be there.

SILBERMAN: *One of the things that has surprised me in meeting autistic adults is how much can be going on inside them that is not apparent from the outside.*

ROBISON: I see that and see the way that people with seemingly severe autism obviously see something in me. If you were me, or even if you were just watching me with them, it's unmistakable. I talked to Temple Grandin about this. I told her, "I'm afraid that when I talk to groups of people with really serious autistic disability, they're going to think I'm a fake autistic person, because I can talk so well." Temple said, "No one with autism is ever going call you a fake autistic person. People with autism are like dogs in the park—they absolutely positively know what's a dog and what's a cat. A German shepherd can come up to a dachshund and he will never mistake a dachshund for a cat." And that turned out to be true for me.

So many times, I've seen it in myself going the other way. All these people come up to me at book signings and stuff and I say, "Ah, I see you're a fellow Aspergian," and they say, "How do you know?" And I say, "You know, I don't know." Sometimes they're insulted, because they worked so hard in school, and they look so good, and they have friends and stuff. They're almost insulted that I picked it out. But I don't always know what it is about them.

There's something that ties us all together like that. Studies show people with autism form language a little bit differently, so the cadence and rhythm of our speech is different. We also operate our motor cortexes differently, so we have a little roll and waddle to our walk. There are all these subtle things that are characteristic of autism that nobody recognized before folks like Temple and me started writing books and getting people to talk about it.

SILBERMAN: *How did you end up becoming one of the first autistic people in an official advisory position with Autism Speaks?*

ROBISON: It happened because I read so much criticism, not just of Autism Speaks, but also of the NIH [National Institutes of Health] and the CDC [Centers for Disease Control and Prevention]. So many people were writing online about how we shouldn't be spending money on this or we shouldn't be doing that. I felt like, "What do you guys want to do? Stand on the street corner holding up protest signs, or do you want to go inside and actually make a difference? I frankly think you're wasting your time protesting on the street and writing about it on a blog. If you want to change how things work, you join the organization and change it." So that's what I decided to do.

SILBERMAN: *There's so much speculation about why the number of autism diagnoses seems to be going up so steeply. Do you think the inclusion of people like you on the spectrum would cause the autism numbers to rise dramatically from what they were twenty-five years ago?*

ROBISON: I'm not even sure that the real numbers of people with autism have increased dramatically from twenty-five years ago. People will tell you, "But now there's a kid with autism in every classroom!" First of all, that's not a true statement. Maybe there's a kid with autism in every three, four, or five classrooms—not in every one. And consider that many kids with significant autism would have fallen under the umbrella of mental retardation in 1910. That umbrella encompassed one in fifty children. The whole autism spectrum—including people with the Asperger's diagnosis, who would not have been diagnosed with mental retardation—might ultimately include one in a hundred. So I don't think that there's any evidence that the numbers have really grown bigger than that.

SILBERMAN: *What are the most promising areas in autism research right now?*

ROBISON: I think TMS [transcranial magnetic stimulation] has tremendous potential to alter speech functionality and emotional intelligence in people. Genetic research may give us tools to prevent the onset of some of the most severe autistic disability in eight or ten years. I think there are some tremendously promising behavioral therapies. The single most promising thing in the world of autism is pushing the age of detection from six or eight back to two. If we do that and apply intervention, a large percentage of those kids will not be disabled as grown-ups. That's the most promising thing being done.

SILBERMAN: *If you could say something right now to a fourteen-year-old who has just been diagnosed, what would you say?*

ROBISON: I would tell them to look at the evidence of my life and the other geeks who are middle-aged. It gets better. If you look at the visible changes in me over four or five years, or the visible changes in Temple Grandin over ten years, we are proof that change and improvement do not stop at age eighteen or twenty-one, but continue for your whole life. There's a very good outlook for young people.

STEVE SILBERMAN is a longtime writer for *Wired* and other national magazines and a blogger for the Public Library of Science. He is writing a book about autism and neurodiversity called *Neurotribes: Thinking Smart About People Who Think Differently* for Avery/Penguin. He lives with his husband, Keith, in San Francisco.

42

THE DODO IS DEAD, LONG LIVE THE DODO!

BRIAN SWITEK

The Dodo, *Didus*, is a bird that inhabits some of the islands of the East Indies. Its history is little known; but if the representation of it be at all just, this is the ugliest and most disgusting of birds, resembling in its appearance one of those bloated and unwieldy persons who by a long course of vicious and gross indulgences are become a libel on the human figure.

—Charlotte Turner Smith, *A Natural History of Birds: Intended Chiefly for Young Persons*, 1807

I hate to say it, but the dodo looked as if it deserved extinction. What other fate could there have been for such a foolish-looking ground pigeon? A grotesque, tubby creature with huge nostrils and a ridiculous little poof of tail feathers, *Raphus cucullatus* had the air of a bird that stood still with a blank stare as the scythe of extinction lopped off its head.

But the dodo I have always known is not a true reflection of the bird. Notes, skeletal scraps, a disregard for soft-tissue anatomy, and a bit of artistic license created this symbol of extinction. The dodo looked so stupid because we made it so.

In order to understand the legacy of the dodo, a little background on its demise is required. It has not been very long since we lost the dodo—only about three centuries—but the exact date has been difficult to pin down. Until recently, the last confirmed dodo sighting on its home island of Mauritius was made in 1662, but a 2003 estimate by David Roberts and Andrew Solow placed the extinction of the bird around 1690. They were probably not far off.

Historical documents described by Julian Hume, David Martill, and Christopher Dewdney in 2004 confirmed that dodos were killed for the Opperhoofd (governor) of Mauritius, Hubert Hugo, on August 16, 1673. Hugo's successor, Isaac Joan Lamotius, also jotted down notes about still-living dodos in his notebooks at least twelve times between 1685 and 1688, with the last capture of a dodo recorded on November 25, 1688. (There is some doubt here, as some historians think Lamotius was referring to the also-extinct red rail, but Hume and coauthors pointed out that Lamotius was a skilled observer of nature who was unlikely to confuse this distinctive dodo with the red rail.)

Using these late sightings with the estimation techniques of Roberts and Solow, the scientists came up with a new extinction date of 1693, although we will probably never know when the last dodo actually died. Over a century before the idea of extinction was accepted, those who extirpated the dodo did not keep detailed records of the bird's decline. That an entire species could disappear simply did not occur to them.

No single cause drove the dodo into the abyss. Humans hunted the naive birds, of course, but the rats, cats, pigs, and other animals that we brought along with us were just as destructive. The extinction of the dodo was not simply a matter of systematic extermination. Our species created a major ecological disruption that many unique island species could not cope with. Still, the fact that dodos were regularly hunted and killed greatly contributed to their demise, and, contrary to the common belief that they had a disgusting flavor, Jan Den Hengst has drawn on several historical sources to show that dodo meat was considered by sailors to be quite palatable. Who knows how many dodos were killed to satisfy gustatory curiosity?

Thankfully for us—though not for the dodo—some of those hungry sailors recorded a few aspects of the bird's natural history. The Dutch who stayed on Mauritius observed the dodos, made notes about them, sketched them, and even brought stuffed dodos back to Europe. What they saw were leggy, light-bodied birds that strutted rather than waddled. So why, then, are there so many inaccurate restorations? We're not dealing with some animal that became extinct during the Pleistocene, living on only in sketches drafted on cave walls. The Age of Exploration both discovered and wiped out the dodo—from a geological perspective, it popped out of existence only yesterday—and it is puzzling why an animal that died out so recently has been so poorly represented.

In many cases, mistakes about the dodo are copycat errors. One artist got something wrong and the mistake stuck. Take the color of the dodos, for example. Firsthand accounts of the birds agreed that they sported plumage that was black to gray in color, but many seventeenth-century Dutch paintings restored them as white. Why they did so is unknown—perhaps artists mistakenly gave the dodos the color of another now-extinct bird, the white ibis of Réunion Island, or perhaps the unique coloring of an albino dodo caused it to be copied more regularly than others. Whatever the reason, light-colored dodos stuck around.

A single painting by Roelandt Savery had an even stronger effect. His rendition of the dodo, created around 1626, differed from earlier drawings of dodos as long-legged and spry in showing the dodo as a fat, stumpy bird. Although earlier illustrations of live dodos had been made by travelers to Mauritius, Savery's was by far the most ornate, stylized, and detailed painting, so it is not surprising that subsequent artists followed his lead. Even Richard Owen, the brilliant Victorian anatomist, later used Savery's rendition as a starting point for reconstructing the bird.

We can't be too critical of Savery, though. There are only two confirmed accounts of live dodos that were displayed in Europe, and Savery probably never saw a still-breathing dodo. Most artists who illustrated the bird never saw a living specimen. This situation left at least one telltale sign in artistic renderings of the flightless avian—the enlarged nostrils. Sketches of live and recently deceased birds show the nostrils as being very small, but in skeletons and stuffed specimens the soft tissue was gone, leaving a ludicrously large nasal cavity. If a dodo restoration has large, gaping nostrils, then it was based upon a long-dead bird.

Mistakes about dodo anatomy gained a cultural inertia that was difficult to stop. Extensively reviewed by dodo expert Julian Hume in 2006, illustrations of dodos were based on scrappy remnants and the works of others. "The Dodo, one of the most documented and famous of birds and a leading contender as the 'icon' of extinction," he wrote, "has endured more than its fare [*sic*] share of over zealous misinterpretation." At least the Pleistocene artists who drew mammoths, rhinos, and Irish elk on cave walls had seen the living creatures; in the case of the more recently extinct dodo, the distance between the artists and the last birds allowed errors to take hold and quickly proliferate.

Strangely, though, the dodo became an almost mythical creature as soon as it became extinct. Samuel Turvey and Anthony Cheke documented in 2008 that, despite the bird's notoriety among the Dutch, many French naturalists considered the bird to be entirely fanciful. To some eighteenth-century naturalists, the dodo was about as real as a griffin, and there seemed to be no conclusive evidence that the bird had ever actually existed. Given that the French took control of Mauritius in 1715 and found no sign of the dodos, it seemed possible that the birds were the product of exaggeration and overactive imaginations.

It was only in the early nineteenth century, when European naturalists began describing dodo scraps scattered among various museums, that it became widely recognized as a real animal that had recently gone extinct at the hands of our species. (And, of course, its appearance as an icon of political foolishness in *Alice's Adventures in Wonderland* helped.) "Both chance and necessity played a part in the Dodo's rise to fame," Turvey and Cheke noted, and the dodo became symbol of extinction only when more recent extinctions—such as that of the great auk in the mid-nineteenth century—affirmed that species truly could undergo a catastrophic decline. Scientists working today know more about the dodo than the naturalists who overlapped in time with the last birds, although much about this strange bird remains uncertain.

Among the frustratingly fuzzy questions about the dodo was how much it weighed. Here the notes of eyewitnesses and estimates made by scientists conflict. Whereas some mariners said that dodos weighed as much as 50 pounds, scientific estimates based upon the bird's skeletal anatomy have restored them at between 23 and 46 pounds. The higher estimate is consistent with the pudgy, waddling creature seen in seventeenth-century paintings, whereas the lower bar fit the earlier reports of svelte, long-legged dodos.

According to a paper published in early 2011 by Delphine Angst, Eric Buffetaut, and Anick Abourachid that used leg bones—from femur to ankle—to estimate the mass of the bird, dodos may have come in at just under the previous lower limit. Dodos weighed only about 22 pounds. This is about as heavy as a wild turkey, and the scientists proposed that the heavier estimates of the seventeenth-century mariners may have been inspired by the puffed-up appearance of some birds, plus a bit of exaggeration.

In order to truly understand the dodo, though, we need more remains of the bird. Despite the number of preserved dodos brought back to Europe, scientists have rarely had the opportunity to study whole skeletons. The scant sampling of dodo remains collected during the seventeenth century were lost, destroyed, and crumbled to dust.

According to a famous bit of natural history apocrypha, by 1755 the last remaining stuffed dodo in Oxford's Ashmolean Museum had degraded to such a point that it was ordered to be destroyed in a fire. It was only through the quick intervention of a sharp-eyed naturalist that the head and foot were saved from the flames. As with many cherished stories, though, this is untrue. The dodo was so badly decayed that the museum's curator had the head and foot removed so that they might be saved from the otherwise rotted mount. Other dodos succumbed to a similar fate. What remained of the birds almost went through a second extinction that would have robbed naturalists of any evidence the avians ever existed. Mercifully, bones and other scraps were spared.

The first scientific assessment of a complete dodo skeleton was made in 1866 by Richard Owen. He had reconstructed the skeleton of the dodo from the subfossil remains of multiple individual birds found on Mauritius, although Owen's vision was controversial for two different reasons. From an anatomical perspective, Owen had assumed that Savery based his painting on a living bird and simply reconstructed the bones to fit within an outline of the artist's plump dodo. (Owen later issued an updated, more lightweight version of the dodo skeleton in 1872.)

Owen's ability to reconstruct the bird at all, though, was made possible by his hijacking of fossils intended for naturalist Alfred Newton at Cambridge. Through a bit of haggling with contacts in London, Owen intercepted shipments of dodo bones meant to go to Newton and additional specimens meant for sale. This coup ensured that only he would have the materials needed to scientifically revive the bird in print. Owen's stake on the dodo forced Newton to reluctantly bare his throat by offering Owen the best of the dodo fossils he had in his possession and also withdrawing his own paper on the dodo from potential publication, letting Owen become the prime interpreter of yet another fantastic extinct creature.

Because of the lack of any stuffed specimens or new skeletons, it is easy to see how the traditional image of the buffoonish dodo remained entrenched, but recent expeditions to Mauritius have turned up new fossils of the bird. A 2007 report stated that the most complete dodo skeleton ever found was recovered from a cave deposit, and a 2009 paper by Kenneth Rijsdijk, Julian Hume, and colleagues described a 4,000-year-old bonebed rich in dodo remains.

This site has provided researchers focused on the not-too-distant past a fleeting glimpse into what Mauritius was like long before the arrival of the Dutch sailors. In addition to numerous dodo remains, the bones of extinct giant tortoises, bats, and other birds were found in the same deposit, which has been reconstructed as a freshwater oasis in an otherwise dry habitat. According to an update on the site published by Rijsdijk and collaborators in 2011, though, this place of refuge dried up sometime between 4,235 and 4,100 years ago. The bones found there—including those of the dodo—represent the desiccated victims of an ancient megadrought. Dodos and the other animals survived in other pockets, but the catastrophe underscores the importance of fresh water to strange and unique creatures restricted to islands. As human-caused climate change continues, and island habitats are destroyed by our activities, other wondrous island-bound species will be lost. The more we learn about the dodo's history, the more the bird becomes a potent symbol of lost species.

Despite its proximity to us in time, it is almost easier to think of the dodo as a fossil creature. So much of what we thought we knew about it relied on the testimony of long-dead witnesses. Only by going back to the bones of the dodo can we start to understand this bird's biology. The dodo is an unmistakable icon of extinction, a species squandered in near time, but separating the animal from its modern mythology is an ongoing task.

REFERENCES

Angst, Delphine, Eric Buffetaut, and Anick Abourachid. 2011. "The End of the Fat Dodo? A New Mass Estimate for *Raphus cucullatus*." *Naturwissenschaften* 98 (3): 233–36. doi:10.1007/s00114-010-0759-7.

Den Hengst, Jan. 2009. "The Dodo and Scientific Fantasies: Durable Myths of a Tough Bird." *Archives of Natural History* 36 (1): 136–45. doi:10.3366/E0260954108000697.

Hume, Julian P. 2006. "The History of the Dodo *Raphus cucullatus* and the Penguin of Mauritius." *Historical Biology* 18 (2): 65–89. doi:10.1080/08912960600639400.

Hume, Julian P., Anthony S. Cheke, and Alastair McOran-Campbell. 2009. "How Owen 'Stole' the Dodo: Academic Rivalry and Disputed Rights to a Newly-Discovered Subfossil Deposit in Nineteenth Century Mauritius." *Historical Biology* 21 (1): 33–49. doi:10.1080/08912960903101868.

Hume, Julian P., Ann Datta, and David M. Martill. 2006. "Unpublished Drawings of the Dodo *Raphus cucullatus* and Notes on Dodo Skin Relics." *Bulletin of the British Ornithologists' Club* 126 (A): 49–54.

Hume, Julian P., David M. Martill, and Christopher Dewdney. 2004. "Palaeobiology: Dutch Diaries and the Demise of the Dodo." *Nature* 429 (6992): 2688. doi:10.1038/nature02688.

Nicholls, H. 2006. "Ornithology: Digging for Dodo." *Nature* 443 (7108): 138–40. doi:10.1038/443138a.

Rijsdijk, Kenneth F., Julian P. Hume, et al. 2009. "Mid-Holocene Vertebrate Bone Concentration-Lagerstätte on Oceanic Island Mauritius Provides a Window into the Ecosystem of the Dodo (*Raphus cucullatus*)." *Quaternary Science Reviews* 28 (1–2): 14–24. doi:10.1016/j.quascirev.2008.09.018.

Rijsdijk, Kenneth F., Jens Zinke, et al. 2011. "Mid-Holocene (4200 kyr BP) Mass Mortalities in Mauritius (Mascarenes): Insular Vertebrates Resilient to Climatic Extremes but Vulnerable to Human Impact." *The Holocene* 21 (8): 1179–94. doi:10.1177/0959683611405236.

Roberts, David L., and Andrew R. Solow. 2003. "Flightless Birds: When Did the Dodo Become Extinct?" *Nature* 426 (6964): 245. doi:10.1038/426245a.

Turvey, Samuel T., and Anthony S. Cheke. 2008. "Dead as a Dodo: The Fortuitous Rise to Fame of an Extinction Icon." *Historical Biology* 20 (2): 149–63. doi:10.1080/08912960802376199.

BRIAN SWITEK is a science writer and fossil fanatic based in Salt Lake City, Utah. He is the author of the blogs *Laelaps* (*Wired* Science) and *Dinosaur Tracking* (Smithsonian.com), as well as the book *Written in Stone: Evolution, the Fossil Record, and Our Place in Nature*. His next book—*My Beloved Brontosaurus: On the Road with Old Bones, New Science, and Our Favorite Dinosaurs*—will be published in 2013. Find him at http://brianswitek.com.

43

SEX AND THE MARRIED NEUROTIC

MELANIE TANNENBAUM

There are few things in this world that I truly loathe. One of those things is the show *Everybody Loves Raymond*.

Why? you might ask.

First of all, it's actually quite hard to really "love" Raymond. From what I've seen of the show, he seems to care about three things: golf, trying (in vain) to have sex with his wife, and placating his intrusive family.

But there's another problem—Debra isn't innocent either.

For those who have never seen the show, Debra—the wife of the eponymous Raymond—manages to embody practically every negative stereotype about middle-aged, suburban, married women with children. She is cold toward her husband, and her moods swing at the drop of a hat: she might cry about how much she loves her husband in one second and then shove him away as she storms upstairs in the next. And she seems interested in having sex only when she can use it as a negotiation tactic or the means to some sort of manipulative end.

Personality psychologists would describe Debra's temperament as highly neurotic—known in everyday parlance as emotionally unstable. Neurotic people are more likely than their non-neurotic counterparts to experience negative emotions (such as anxiety or depression) and frequent mood swings, and would likely agree strongly with personality questions like "I change my mood a lot" or "I get irritated easily."

Unfortunately, neuroticism isn't cute. It's not an adorable personality trait that we can laugh off because it's funny that "women are crazy." First of all, the sexism inherent in that assumption is cringe-inducing. And second, neuroticism is particularly bad for your marriage. In fact, neuroticism is the one personality trait that best predicts marital dissatisfaction, separation, and divorce. If you want to know whether a couple will still be together in ten years, you might want to start by looking at how often both partners feel irritable or experience mood swings.

Logically, this makes sense. Neuroticism, by definition, makes a person more likely to experience negative emotions. If someone is prone to feeling sad, anxious, or irritable, this person will most likely also feel sad, anxious, or irritable about his/her relationship—and this person's partner will likely feel less satisfied as well. After all, it's easier to be happy when you're around a happy person.

But there's a catch.

Think now about Phil and Claire Dunphy, one of the couples from the current sitcom *Modern Family*. For those not familiar with the show, Claire is also quite neurotic; she is prone to overwhelming amounts of anxiety and has frequent bursts of extreme irritability. Yet the Dunphys generally seem much more satisfied with their marriage than Ray and Debra Barone. So what's the difference? Both couples have been married for twenty-plus years. They're both middle-aged. They both have three children.

Well, first read this quote from the *Everybody Loves Raymond* Wikipedia page: "Running Gags: As in many episodes, Ray tries to have sex with Debra, but something inevitably prevents it."

Now, contrast this depiction of Ray and Debra's sexual relationship with a scene from *Modern Family* in which the Dunphys are still reeling in humiliation from their three children walking in on them "doing the deed."

Phil and Claire have sex.

Regularly.

They've been married for twenty years, yet they still have excited, we-still-really-want-each-other, dress-up-as-strangers-and-pick-each-other-up-in-the-hotel-bar-on-Valentine's-Day sex.

According to recent research, sex might be the golden ticket. After newlywed couples were surveyed several times during the first four years of their marriages, the couples who didn't have sex frequently showed the same old pattern—the more neuroticism there was in the marriage, the more dissatisfied the partners were.

However, for the couples who frequently had sex, neuroticism didn't seem to matter at all. In fact, when neurotic couples were having a lot of sex, they were exactly as satisfied with their marriages as the non-neurotic couples.

You may want to argue that studying newlyweds does not say much about what that couple would be like after twenty years. However, in the original studies, neuroticism levels predicted only initial satisfaction, which in turn predicted later satisfaction and/or divorce. Neuroticism does not predict anything about how a marriage will change over time, the likelihood that an initially happy couple may become unhappy, or the different rates at which partners might reach "unhappiness." It's safe to say that any impact of neuroticism on marital dissatisfaction in the original studies was also there when the couples were newlyweds.

Of course, this finding begs some questions. First of all, the authors don't specify exactly what "a lot of sex" is. However, if they used the fairly conventional approach of forming categories based on the overall mean and standard deviation, they may have defined "low frequency" as two times per month (or less frequently) at the very beginning, and one time every two months (or less frequently) by the fourth year of marriage. "High frequency," on the other hand, was likely defined as around fourteen times per month (or more) at the very beginning, and twelve times per month (or more) by the fourth year of marriage. But people are individual creatures, with different sex drives, wants, and needs, and each person probably has a different idea of what "a lot" or "a little" actually means. While fourteen times per month might seem woefully inadequate for one person, it might be far too often for another. So what matters more—the actual number of times that you have sex per month or feeling like you have it often enough?

Second, the authors make sure to control for a lot of possible objections. The time-lag design that they used convincingly rules out the possibility that being happier in the first place leads couples to have more sex as a result, and the link between sex frequency and satisfaction in neurotic couples remains even after they control for several important variables (such as gender, the quality of "nonsexual" relationship domains such as trust/communication/affection, and attachment insecurity). It truly seems that the frequency of sex is what leads neurotic couples to feel happier—not the other way around, and not due to a theoretically plausible third variable.

But if this is true, the really compelling question is why? Why does sex make emotionally unstable partners more satisfied with their relationships? Does sex create a greater sense of intimacy between relationship partners, leading them to feel more "stable"? Is it purely neurochemical, with the frequent dopamine rushes from sex making these partners consistently happier? After all, sexual behavior one day does lead to fewer negative moods the next day. Are any of these differences in self-reported marital satisfaction simply biased responses based on perceived cultural norms ("We have sex more often than my friends have sex with their husbands, so we must be happy")?

Here's another intriguing twist: sex frequency didn't make any difference in the marital satisfaction of the non-neurotic couples—it only brought the previously less happy neurotic couples up to the non-neurotic couples' satisfaction level. So it doesn't seem to be a general effect of sex making married couples happier—it specifically impacts neurotic couples.

This makes me wonder: Is there something about being "emotionally unstable" that leads neurotic people to actually get more out of sex than those who are more stable? Does having a brain wired for strong emotionality make neurotic people respond to sex in a way that's simply different from others? Is sex an effective physical outlet for the extreme emotionality that accompanies neuroticism, such that people who experience erratic mood swings can release these strong emotions through sex instead of taking them out on their loved ones?

We can't know the hows or whys quite yet. All I can say is that Debra Barone's tactic of withholding sex when she was upset might not have been as good for her marriage as she thought.

REFERENCES

Costa, Paul T., and Robert R. McCrae. "Normal Personality Assessment in Clinical Practice: The NEO Personality Inventory." *Psychological Assessment* 4, no. 1 (1992): 5–13. doi:10.1037//1040-3590.4.1.5.

Karney, Benjamin R., and Thomas N. Bradbury. "The Longitudinal Course of Marital Quality and Stability: A Review of Theory, Method, and Research." *Psychological Bulletin* 118, no. 1 (1995): 3–34. PMID: 7644604.

——. "Neuroticism, Marital Interaction, and the Trajectory of Marital Satisfaction."

Journal of Personality and Social Psychology 72, no. 5 (1997): 1075–92. doi: 10.1037//0022-3514.72.5.1075.

Russell, V. Michelle, and James K. McNulty. "Frequent Sex Protects Intimates from the Negative Implications of Their Neuroticism." *Social Psychological and Personality Science* 2, no. 2 (2011): 220–27. doi:10.1177/1948550610387162.

Burleson, Mary H., Wenda R. Trevathan, and Michael Todd. "In the Mood for Love or Vice Versa? Exploring the Relations Among Sexual Activity, Physical Affection, Affect, and Stress in the Daily Lives of Mid-Aged Women." *Archives of Sexual Behavior* 36, no. 3 (2007): 357–68. doi:10.1007/s10508-006-9071-1.

MELANIE TANNENBAUM holds a B.A. with honors in psychology from Duke University and an M.A. from the University of Illinois at Urbana-Champaign, where she is currently a doctoral candidate in social psychology. She blogs at *PsySociety* (http://psysociety.wordpress.com) and has contributed to *Scientific American* and *In-Mind* magazine. More information about Melanie and her writing can be found at her personal website, www.melanietannenbaum.com.

44

WHY DO WOMEN CRY? OBVIOUSLY IT'S SO THEY DON'T GET LAID

CHRISTIE WILCOX

The headline alone was enough to make me gag: "Stop the Waterworks, Ladies. Crying Chicks Aren't Sexy." The sarcastic bitch in me just couldn't help but think, *Why, THANK YOU! I've been going about this all wrong. When I want to get some from my honey, I focus all my thoughts on my dead dog or my great-grandma and cry as hard as I can. No WONDER it isn't working!*

I didn't even want to read the rest of the article by Brian Alexander, who writes the Sexploration column for MSNBC.

But I did.

It doesn't get better.

Alexander's reporting of the actual science underlying a recent study published in *Science*—about men's physiological responses to the chemicals present in women's tears—was quick and simplistic, and couched in sexist commentary (like how powerful women's tears are as manipulative devices). And to finish things off, he clearly states what he found to be the most important find of the study: "Bottom line, ladies? If you're looking for arousal, don't turn on the waterworks."

It's no wonder that the general public sometimes questions whether science is important. If that was truly the aim of this paper, I'd be concerned, too!

I don't think Brian Alexander is a bad guy or a misogynist. Sure, his job is all about selling sex stories to the public. He even wrote a book about American sexuality. But I don't personally think he has a burning hatred for women or views them as objects placed on this earth for the sexual satisfaction of men.

(Courtesy of Sara LeeAnn Banevedes)

Brian Alexander just missed the point. This paper wasn't published as a part of a women's how-to guide for getting laid. Instead, the authors sought to determine if the chemicals present in human tears might serve as chemosignals like they do for other animals—and they got some pretty interesting results.

In many species, chemical signals run rampant. Scents, pheromones, and other chemical cues are deliberately and unconsciously given off to tell other individuals anything from "Back off—*my* tree!" to "Hop on and ride me, baby!" But despite how common they are in the rest of the animal kingdom, the function of chemical signals in humans is hotly debated. After years of searching, we have yet to find human pheromones (no matter what those websites tell you), and while scent seems to play a role in communication in people, there is still relatively little knowledge as to what chemicals and why.

Given that tears are known to serve as sexual signals in mice, it isn't strange at all that Noam Sobel and his team chose to look at the physiological responses to tears. The Israeli team designed an impressive and unbiased set of experiments to determine if the tears produced by women when sad elicit physiological responses in men separate from the visual or auditory stimuli of a woman crying.

To find out if tears alone act as chemosignals, the scientists collected tears from women watching tear-jerkers, and as a control, compared their effects to saline rolled down women's cheeks. Men sniffed the solutions without any knowledge of what they were during a series of different experiments. In the first, men with a tear-soaked pad under their nose were asked to rate the sexual attractiveness and mood of female faces. While the smell of saline had no effect, men inhaling Eau de Tears consistently rated women's faces as less attractive, though this had no impact on whether they found the faces happy or sad.

For the second experiment, men sniffed tears before watching a sad movie. While doing so didn't affect their mood, the smell of tears did elicit a physiological response: men's faces became more conductive to electricity, which happens when we sweat and is indicative of a psychological reaction. Furthermore, the men self-reported less sexual arousal, which was reflected in their bodies as a 13 percent drop in saliva testosterone levels.

But to really get to the meat of it, the team threw their male test subjects into an fMRI machine and scanned their brains for activity while sniffing tears. Researchers saw much less activity in the hypothalamus and the fusiform gyrus, both of which are thought to be involved in sexual arousal. All three experiments lead to the same conclusion: the chemicals in women's emotional tears reduce male sex drive.

The real question, though, is why. Why do men's testosterone levels tank at the smell of a woman's tears? The overwhelming answer given by mainstream media is that tears just aren't sexy. When women cry, so the journalists say, it's a chemical signal that they don't want to have sex. Because evolution is all and only about sex . . . right?

Sorry to burst their bubble, but even when it comes to evolution, it's not all about sex. Selection also favors survival—because, you know, you can't have sex when you're dead. Thus women's tears are not necessarily evolutionarily intended to turn guys off. For example, there's a hypothesis that tears might be used to dampen aggression. Think about it: we cry when we're sad or physically in pain. In both cases, we're more vulnerable. Getting others, especially angry men, to be less aggressive toward us in that moment could certainly be a benefit to survival.

Really, the idea that tears are intentionally used as a turn-off is a hard sell to an evolutionary biologist. What benefit do women get from not having sex when crying? Does it somehow make them have healthier or more babies? Not for any reason I can think of.

There is, instead, an even more intriguing explanation, one that makes a whole lot more sense. Many who wrote about this paper (including Brian Alexander) mentioned that tears are known to contain a variety of compounds, including prolactin, the hormone that is responsible for making a guy cool his jets after he gets off. But prolactin does much more than ensure a guy stops going at it—it's a hugely important hormone for nurturing behaviors. In fact, the connection between reduced testosterone and nurturing/bonding behaviors may be the real reason why men's testosterone levels dip upon sniffing tears.

Numerous studies have shown that parental and nurturing behaviors are mediated by prolactin while inhibited by testosterone. For example, research has shown that prolactin levels positively and testosterone levels negatively correlate with a father's impulse to respond to a baby's cry. Furthermore, men's prolactin levels spike and testosterone levels drop in the weeks before their partner gives birth.

It goes beyond babies, too. Decreased testosterone and increased prolactin are strongly implicated in establishing and maintaining relationships. Monogamous men have significantly lower testosterone levels and higher prolactin levels than their single brethren. Furthermore, studies have directly shown that artificially increasing testosterone in a double-blind, placebo-controlled setting makes men less generous to strangers and reduces a person's empathy for others.

Perhaps prolactin or other chemical signals in tears are directly targeting and activating the nurturing pathway in men's brains. Being taken care of or protected when in emotional or physical pain would definitely benefit an individual's survival. Personally, I would like to see this study of tears replicated to determine women's responses to the scent, as well as men's reactions when using men's and children's tears; I'd also like to look at the levels of prolactin, oxytocin, and other well-established bonding and empathetic hormones. My bet is the response isn't limited to men, and isn't limited to emotional secretions from women.

While Brian Alexander and the rest of the sensationalists seem to suggest the signal is "I'm not in the mood," it's likely that the message has nothing to do with having or not having sex. Women aren't saying "back off"—they're saying "help me."

Why do I care so much? It's not just that they got it wrong. It's that their interpretation of research isn't labeled as opinion. It's that the vast majority of people who have any interest in science news are going to read inaccurate (if not downright insulting) news articles and think studies like this one are either misogynistic or frivolous. It's that dumb-asses like Brian Alexander undermine good science for the sake of attention-grabbing headlines. As a scientist and a writer, I am doubly insulted.

REFERENCES

Alexander, Brian. "Stop the Waterworks, Ladies. Crying Chicks Aren't Sexy." *The Body Odd,* January 6, 2011. bodyodd.msnbc.msn.com/_news/2011/01/06/5772347-stop-the-waterworks-ladies-crying-chicks-arent-sexy.

Gelstein, Shani, Yaara Yeshurun, Liron Rozenkrantz, Sagit Shushan, Idan Frumin, Yehudah Roth, and Noam Sobel. "Human Tears Contain a Chemosignal." *Science* 331, no. 6014 (2011): 226–30. doi:10.1126/science.1198331.

CHRISTIE WILCOX is a science writer who moonlights as a Ph.D. student in cell and molecular biology at the University of Hawaii. She hopes to continue her dual-sided career after she obtains her degree, serving as a direct link to the field she loves. Find more about her life and work by visiting her blog at ScientificAmerican.com (http://blogs.scientificamerican.com/science-sushi) or her website (http://christiewilcox.com).

45

HOW TO TAKE THE REAL MEASURE OF A MAN

ALLIE WILKINSON

Men, I know you are fascinated with your penises. You won't admit it, but you're also fascinated with the penises of those around you. "*What?*" you say indignantly. Don't try to deny it. We know about the furtive glances in the locker room and at the urinal. But we get it—it's your way of sussing out the competition, seeing how you measure up. Forget about your penis for a moment. Instead, let's talk about that space between your testicles and your anus, a.k.a. the taint ('taint anus, 'taint balls). I'm going to let you in on a little secret—the length of your member doesn't really have a whole lot to do with your reproductive fitness, but the length of your taint does. So it's time to put away the rulers and whip out the digital calipers!

A new study came out in *PLoS ONE*, claiming to be the first assessment of anogenital distance in adult men—i.e., the length of the taint—as well as the first examination of the relationship between anogenital distance and a man's fertility. I hate to break it to the researchers from Baylor College of Medicine and Stanford University School of Medicine, but you are not the first. A *very* similar study came out two months before in *Environmental Health Perspectives*.

Anogenital distance is used to determine the gender of animals—shorter for females, longer for males. It has also been studied in human infants, mirroring what we know to be true in the animal world (which makes sense, because we are animals). But no one had studied this in adults up until recently, which leads us back to the sudden scientific interest in measuring men's taints: "The Relationship Between Anogenital Distance, Fatherhood, and Fertility in Adult Men."

Why do we care how long a man's taint is, anyway? Well, because in the past half century there has been a reported decline in semen quality and male births, with an increased rate in male genital abnormalities and testicular cancer—and no one likes cancer or malformed genitals or weak semen.

Previous studies have shown that in addition to reduced anogenital distance, exposure to special chemicals added to plastics to increase flexibility and durability, called phthalates, had altered testicular size and the function of Sertoli cells, which nurture the developing sperm cells. Not only that, rodents exposed to endocrine disruptors during critical gestational windows for genital development saw irreparable alterations to penis length, anogenital distance, and testicular weight.

The latest study on the length of the taint, while similar to the earlier one in *Environmental Health Perspectives*, used a slightly different methodology. Eligible patients were recruited from a urology clinic specializing in reproductive medicine. Patients evaluated for infertility who were over the age of eighteen were eligible for the study, and a fertile control group was put together with a group of men with a prior history of paternity.

In order to measure their genitals, the men had to get into a frog-legged position. Oh yes, there are pictures—Note to self: Do not read papers on anogenital distance and male fertility while sitting in a busy airport. You will unknowingly stumble upon said diagram while surrounded by people.

How does one get ready for measurement, anyway? First you lie on your back, and then touch the soles of your feet together, making sure to keep your feet twelve to eighteen inches away from your butt. Oh, and you may want to get a Brazilian wax first. Someone is going to have to get all up in your business, and I'm sure they'd appreciate a clean work surface.

Once ready for measuring, someone will take a pair of digital calipers, and measure from the back of your balls to your anus, which these researchers thought to be a more comfortable, reliable, and reproducible measurement than measuring from the base of the penis. Penis length was taken as well, and testicular volume, or how big the balls are. (I think the pressing question is: *does* Brian Johnson of AC/DC really have the biggest balls of all?) These measurements were estimated by one examiner (lucky man) who had the room at a balmy 78° to 80° Fahrenheit. That temperature is important, because, well, thanks to *Seinfeld*, we all know what happens to the boys when it's cold out:

ELAINE: It shrinks?
JERRY: Like a frightened turtle.

As part of the researchers' routine practice, patients have at least two semen analyses performed, and semen analyses were performed manually within an hour of collection. I'm going to assume that the semen samples were also *obtained* manually, and not with a hands-free, semen-collecting robot.

The volume, density (millions of little swimmers per milliliter), and motility (how fast those suckers could swim) were recorded and multiplied to determine the total motile sperm count. Hormone assays were also processed for testosterone, luteinizing hormone, and follicle-stimulating hormone (FSH)—all of which are hormones that influence fertility in both men and women. Finally, statistical analyses were performed to interpret the data, and a second model was adjusted to account for age, race, FSH, and body mass index.

Most of the men in the study were white males (64 percent), but even when the results were broken down by race, the length-o'-taint differences between fathers and infertile men remained stable. Infertile, childless men had a significantly shorter anogenital distance when compared to fathers. Infertile men also had shorter stretched penis lengths and smaller total testicular volumes than fertile men. All genital measurements seem to be correlated to one another, but those in the know can't stress enough: *correlation does not equal causation*.

Semen volume was similar in both fertile and infertile men, but the sperm density, motility, and total mobile sperm count were significantly lower for infertile men. In both the unadjusted model and the model that accounted for differences in demographics and reproductive variables, anogenital distance and testicular volume were connected with total motile sperm count and sperm density. This connection is considered statistically significant, which means that it is unlikely that these results occurred by chance. Sperm density and total motile sperm count increased with an increase in anogenital distance—for every 1 centimeter increase in taint, add another 4.3 million little swimmers per milliliter and the total motile sperm count increases by 6 million.

There wasn't any significant correlation seen between penis length and sperm count, which goes back to my opening statement: *put away your rulers!* Enough with the size fixation already. Women will not judge your marriage and fatherhood potential on the length of your penis. Instead, they will base it upon your intelligence, your wit, your dashing good looks—and the length of your taint.

REFERENCES

Eisenberg, Michael L., Michael H. Hsieh, Rustin Chanc Walters, Ross Krasnow, and Larry I. Lipshultz. "The Relationship Between Anogenital Distance, Fatherhood, and Fertility in Adult Men." *PLoS ONE* 6, no. 5 (2011): e18973. doi:10.1371/journal.pone.0018973.

Mendiola, Jaime, Richard W. Stahlhut, Niels Jørgensen, Fan Liu, and Shanna H. Swan. "Shorter Anogenital Distance Predicts Poorer Semen Quality in Young Men in Rochester, New York." *Environmental Health Perspectives* 119, no. 7 (2011): 958–63. doi:10.1289/ehp.1103421.

ALLIE WILKINSON (http://alliewilkinson.com) is a freelance science writer and multimedia specialist with a background in environmental studies and conservation biology. Her work has appeared in *Ars Technica*, *Scientific American*, and *Chemical & Engineering News*. She also blogs at *Oh, for the Love of Science!* (http://ohforthelove ofscience.com).

46

THE ORIGIN AND EXTINCTION OF SPECIES

DAVID WINTER

People often ask me, "What do you do?" The shortest answer I can provide is, "I use genetic tools to study evolution." I guess that makes me an evolutionary geneticist. Occasionally I've caught myself saying, "I'm interested in the questions, not the animals." Paraphrased, that becomes something like, "Oh sure, I study Pacific land snails, but for all I care they're just little bags of genes that help me answer questions."

But that's a lie. You can't work on animals without having them affect you. When I started my Ph.D. I had no particular love of snails, but now I'm a complete snail fanboy, and I frequently find myself preaching on the wonders of life as a terrestrial mollusk to people whose only mistake was to ask me what I do for a living. Did you know most slugs retain the remnants of their shells? Or that almost all snail shells coil to the right? Or that mating in many land snail species proceeds only after one snail has stabbed the other with a "love dart"? And then there's the sad tale of the Society Islands partulids.

Believe it or not, land snails are one of the characteristic animals of Pacific Islands. Anak Krakatau is so young it's still smoldering and it has a native land snail species, and Rapa Nui (Easter Island), which is arguably the most isolated island in the Pacific, had its own land snail fauna back when it had forests. It's not entirely clear how these unlikely colonists get to islands. Darwin was so interested in the question he stuck snails to ducks' feet to see if they'd survive an interisland journey. Birds have been shown to carry snails great distances, but windblown leaves are probably a more common mode of conveyance.

We might not know exactly how snails get to islands, but we know what happens once they establish themselves. The land snails of the

Pacific include some of the most outrageous explosions of diversity in the biological world. Chief among these evolutionary radiations were the partulid snails of the Society Islands (the French Polynesian archipelago that includes Tahiti). Partulids are very elegant tree snails that form part of the land snail fauna across most of Polynesia. In total, the tiny islands had sixty-one species of these snails, and each of the main islands has its own endemic forms.

The first person to seriously take up their study was the American embryologist and evolutionary biologist Henry Crampton. Crampton was working at the turn of the twentieth century, a time in which the mechanisms underlying genetics and evolution were very much up for debate, and he hoped Tahitian and Moorean partulids could help set the story straight. Crampton's monographs are famous (at least among people who spend their lives thinking about snails) for their detail. He collected and measured over 200,000 shells, then calculated summary statistics for each species, each site, and each measurement. By hand. To eight decimal places.

Those massive tables might seem like an old-fashioned, descriptive way to do biology. But in many ways Crampton was ahead of his time. For one, he was a Darwinist when not every evolutionist was. By the end of the nineteenth century Darwin had convinced the world of the fact evolution had happened, but relatively few naturalists bought his theory of how evolutionary change happened.

The anti-Darwinian theories that prospered during the so-called eclipse of Darwinism placed very little importance on the variation within species. The orthogenesists and the Lamarckians thought evolution had a driving force, pushing species toward perfection. In their scheme variation within a species was deviance from the mainstream of evolution and was quickly stamped out by natural selection (which they didn't deny, they just said it couldn't be a creative force). Similarly, saltationists thought large-scale evolutionary changes occurred in a single generation, and the small changes you see in populations were of no consequence in the grand scheme of evolution.

Crampton realized that, in a Darwinian world, variation within populations was the raw material of evolution. He was obsessive about measuring his shells because he knew he could use the data he was recording to understand where species came from. In particular, he was able to show that isolated populations of the same species varied from one another. That finding makes sense in light of Darwin's theory, since species arise from populations evolving away from one another; but is harder to fit into progressive theories of evolution, in which you'd expect different populations of the same species to follow the same trajectory.

Crampton's results influenced people like Dobzhansky, Mayr, and Huxley who helped to re-establish Darwinism as the principal theory of evolution in the modern evolutionary synthesis. But Crampton also predicted arguably the most important development in evolutionary theory since the modern synthesis.

In the middle of the twentieth century, evolutionary genetics was defined by a single debate. The "classical" school held that populations in the wild would have almost no genetic variation, because for every gene there would be one "best" version, and every member of the population would have two copies of that gene. Arguing against the classical school, the "balance" school argued that, quite often, there would be no single best gene and organisms would do better having two different versions of the same gene. The balancers thought natural selection would keep lots of different versions of maybe 10 percent of a species' genes.

Both schools assumed natural selection was such a pervasive force that selection would dictate the way populations were made up; they just disagreed on what would result from it. They were both spectacularly wrong. When scientists started being able to measure the genetic diversity of populations in the 1960s, it became clear that almost every single gene had multiple different versions. Now, in the post-genomic age, there is a database with 30 million examples of one sort of genetic variant among humans.

Faced with the overwhelming variation he recorded in partulid shells, Crampton had argued natural selection didn't have a damn thing to do with it. Snails isolated from each other by a mountain weren't adapting to their local habitat, they just varied with respect to traits that had no influence on their survival. The fact that two populations were isolated meant each would follow its own path, and two populations could drift apart from each other.

In light of overwhelming evidence of genetic variation coming from studies in the 1960s, Motoo Kimura proposed the neutral theory of molecular evolution. Kimura's explanation was the same as Crampton's: almost all of the variation we see at the genetic level has no bearing on the success or failure of organisms, so the frequency of different variants drifts around at random. The neutral theory is at the heart of a lot of modern evolutionary genetics, and Crampton had understood the underlying principle fifty years before we knew we needed it.

At the end of his monograph on the partulids of Moorea, Crampton said he'd got as far as his measurements could take him, and it was time for someone to study their genetics. In took a bit longer than Crampton might have hoped, but in the 1960s two leading geneticists took up the study of his snails. James Murray from Virginia and Bryan Clarke from Nottingham spent almost twenty years working in what they called, in more than one paper, the perfect "museum and laboratory" in which to study the origin of species. Their work helped scientists understand, among other things, how ecology can contribute to the formation of new species and what happens to species when they hybridize with others from time to time.

Then, in 1984, Murray and Clarke had to write the most heartbreaking scientific paper I've ever read. It's written in the careful prose scientists use to talk to one another, but the message it delivered was devastating:

> In an attempt to control the numbers of the giant African snail, *Achatina fulica*, which is an agricultural pest, a carnivorous snail, *Euglandina rosea*, has been introduced into Moorea. It is spreading across the island at the rate of about 1.2 km per year, eliminating the endemic *Partula*. One species is already extinct in the wild; and extrapolating the rate of spread of *Euglandina*, it is expected that all the remaining taxa (possibly excepting *P. exigua*) will be eliminated by 1986–1987.

Euglandina rosea is better known as the rosy wolf snail. It senses the mucous trails of other snails, tracks them down, and eats them. It's not clear if the wolf snail had any effect on the pest species it was introduced to control, but it had a huge impact on the partulids. By the time Murray and Clarke wrote their paper, *E. rosea* had already done for one species, and it was too well established to control. All they could do was watch as human stupidity and molluskan hunger slowly (1.2 km, three-quarters of a mile, per year) destroyed the species they'd been studying for twenty years and to which Crampton had dedicated fifty years of his life.

The same slow torture played itself out in Tahiti and then in the rest of the Society Islands. Where there were once sixty-one named species, there are now five alive in the wild. Crampton's hundreds of pages of tables should have been the starting point from which the evolution of the partulids could have been tracked. Murray and Clarke's natural laboratory should still be open and be taking advantage of a new generation of technologies that might be able to reveal the genetic and genomic changes that occur when a new species arises. Extinction is a natural part of life, and the fate of all species eventually, but when it's driven by human shortsightedness and robs us not just of a wonderful product of nature but a window through which we might have understood nature's working, it's very hard to write about.

There is a tiny scrap of good news in this story. The partulids are no longer an iconic genus in the study of evolution, but they have become the pandas of invertebrate conservation. Murray and Clarke were able to get fifteen of the species off the islands and into zoos and labs across the Northern Hemisphere. Breeding programs have been successful, and new lab-based studies come out from time to time. The relict populations back in the Societies don't have nearly the range they used to, but it appears they've held on to most of their original genetic variation. Perhaps, one day, *Eulglandina* can be taken care of and some of the partulids can have their islands back.

REFERENCES

Clarke, Bryan, James Murray, and Michael S. Johnson. "The Extinction of Endemic Species by a Program of Biological Control." *Pacific Science* 38, no. 2 (1984): 97–104.

Crampton, Henry E. "Studies on the Variation, Distribution, and Evolution of the Genus *Partula*: The Species Inhabiting Moorea." *Carnegie Institution of Washington Publications* 410 (1932): 1–335.

DAVID WINTER is a researcher at the University of Otago in Dunedin, New Zealand. He spends most of his time trying to work out what species are, where they come from, and how we know when we are looking at one. When he's not struggling with those questions, he writes at his blog *The Atavism*: www.sciblogs.co.nz/the-atavism.

47

GENOME SEQUENCING AND ASSEMBLY, SHAKESPEARE STYLE

RICHARD F. WINTLE

Our "genome" is the DNA in the cells of our body. It spends most of its time as an unruly-looking blob in the nucleus of the cell but packages itself up nicely into chromosomes when cells divide. It's the genetic code, the material of heredity that passes on traits from parents to children.

The science of genomics, which is what I spend much of my time thinking about, seeks to make sense of the 3 billion or so letters of the genetic code that is written in this DNA. It's helpful to think of it as text—DNA is a long, thin molecule that is made up of four different "letters."

Imagine a string, strung with four types of beads. Each has a single letter on it, and they're all mixed up together. These make a one-letter shorthand for the names of the chemical units that make up DNA: adenine (A), cytosine (C), guanine (G), and thymine (T). When genome scientists talk about "reading the DNA sequence," this is all they mean: what is the order of those beads on the string? We use very sophisticated equipment to read it, but really, that's all it comes down to in the end.

DNA sequence is terribly boring to look at. Here's an example—in this case, a piece of a gene that is responsible for making salivary amylase, an enzyme that digests sugars in your food:

TGGTATCTGTACATACCTTTGATGTCAGTGTTTAG
TACACGTGGCTTGGTCACTTCATGGCTAA

Doesn't look like much, does it? Now, imagine 3 billion letters of this, arranged in forty-six enormous volumes. Those volumes are chromosomes; most people have one each of chromosomes 1 through 22, and two X chromosomes if they're female, an X and a Y if male. That 3 billion letters is roughly equivalent to 857,000 pages of text, or about 28,000 copies of a medium-sized Shakespeare play (say, *Romeo and Juliet*).

The problem of understanding the genome is that while Shakespeare is written in a language that we understand, using familiar concepts (love, jealousy, betrayal) and words that we can look up in a dictionary, the genome sequence is not. It's a featureless plain of those four letters. It's got a great deal of meaning embedded in it, though, and much has been done to understand it.

While a lot of that information came from complicated, specialized biology, some can be found by comparing one genome sequence to another—in other words, looking at variability between individual people. Just as our outward anatomy (hair and eye color, height, the shapes of our noses) varies from person to person, so does the genome sequence. So how do we find this variation?

Returning to Shakespeare, suppose we have a modern edition of *Romeo and Juliet*, and we suspect that it might have some typographical errors in it. To find them, we could compare it to a "gold standard"—perhaps the first printed edition, or better yet, one of Shakespeare's original manuscripts. By comparing the language, we could find errors that change the meaning. Some of them will be obvious. Here's a very famous line from Act II, Scene 2:

> O Romeo, Romeo! wherefore art thou Rodeo?

You don't need to compare with the original, or even to know the play, to infer that there's probably an error in the last word. Genomics researchers can do the same thing—if you show me part of a gene's sequence, I might be able to guess that one of those A, C, G, or T changes is a problem. That comes with experience, just like reading and speaking English provides you with the experience to guess that "Rodeo" should read "Romeo."

Reading the rest of the play would make you even more confident that it's a typo—there are no references to "rodeos" anywhere else in its nearly 26,000 words. Genome scientists use this approach, too, relying on computer programs to find things that just "don't belong." Rather than rodeos in Shakespeare, we look for changes in DNA that just don't occur much (like a "STOP" signal in the middle of a gene). Even without knowing what that gene is supposed to look like, we might infer that such a genetic typo would be bad.

Other errors might be a lot tougher to spot. Consider this quotation, from right after the first one:

> Deny the father and refuse thy name;

Without knowing the play, you'd never be able to guess there's an error there—the first "the" is supposed to read "thy." It's just one little letter that changes the meaning a bit, but it's hard to spot because either "the" or "thy" makes sense. To find it, you need that "gold standard" to compare with.

This is essentially the same as sequencing my genome and comparing it to yours. They're both editions of the same book, and tiny differences can have impacts that are huge (a mutation that makes me sick), modest (a change that gives me a higher risk of getting sick), or inconsequential. Recent studies suggest that among the 3 billion or so letters of our genomes, each of us differs by something like 3 million single-letter typos, and another 45 million that are rearranged in big chunks (in the wrong place, in the wrong order, duplicated, or completely missing). Fortunately, almost all of these don't seem to have much impact on our health.

Our "gold standard" Shakespeare script is likely to have been pieced together from at least five different Quartos and Folios, which is also how the first human genome reference sequence was made. This reference was assembled from sequences of DNA from nearly 750 different sources. It's still extremely useful, but it's only recently that complete sequences from individual humans have become available instead. And just as we use annotations in the margins to tell us what Shakespeare meant by "in choler," or how one might go about getting hoisted with a "petard," so also do genome scientists use annotations to describe different pieces of that 3-billion-character book—where the genes are, for example.

So genomes are like Shakespeare, and variations between people are like typographical errors. Sometimes they're invisible (suppose I switched the places of the two letters *o* in the word "too"), sometimes they don't change the meaning much ("the" and "thy"), and sometimes they're disastrous (where *is* that rodeo, anyway?). Using modern genome science, we can find them, if, as Romeo says, we "know the letters and the language."

> Once more unto the breach, dear friends, once more;
>
> —*The Life of King Henry the Fifth*, Act III, Scene 1

Although I haven't used the term till now, the process of finding a short DNA sequence in the magnum opus that is our genome is usually referred to as "mapping." If you find out where it came from, you've "mapped" it to its proper genomic location. By analogy, we can take an isolated phrase (like the famous "Romeo, Romeo, wherefore are thou . . ." speech) and locate it by comparing with the whole text of the play. Doing this, we'd find it near the beginning of Act II, Scene 2.

That works well for organisms that have had their genomes sequenced, like humans, mice, cattle, or bacteria—hundreds of species. But what about the millions of species with no available genome sequence?

Imagine that you are a researcher who has just discovered a brand-new species of beetle (there are lots of species of beetles, so this is actually pretty likely). You might want to see what's in its DNA that makes it different from other beetles, or find some genetic evidence to help work out how it's related to other species.

Most DNA sequencing specialists would take the approach of extracting the poor beetle's DNA, fragmenting it into manageable pieces, and sequencing these small fragments en masse. There are a number of good technologies for doing this, and they all give you the same result: millions and millions of short "words," made up of the letters A, C, G and T, the four chemical bases that make up DNA. These words are randomly derived from all over the hapless insect's genome. Putting them together to make the beetle version of that 857,000-page book (beetle genomes are generally a bit smaller—so maybe 43,000 pages) is like doing a huge jigsaw puzzle. One with really tiny pieces, all of more or less the same color—and without the picture on the box lid to help you.

This process, called assembly, relies on finding overlaps between sequences. Remember, these are all random fragments, starting and ending at different places, so if we have enough DNA (and thus enough copies of the genome) to begin with, by chance some of these fragments will have recognizable places where they overlap. And this is where *Julius Caesar* comes in.

Imagine that you've never read the play. Think of this set of text fragments as being like a handful of short DNA sequences:

```
ds, Romans, count
ns, countrymen, le
Friends, Rom
end me your ears;
trymen, lend me
```

If you didn't know the play, you might have trouble making sense of this—until you realize that some of the fragments overlap. When you shuffle them around and line them up, you get something like this:

```
Friends, Rom
     ds, Romans, count
                ns, countrymen, le
                        trymen, lend me
                                end me your ears;
```

Which gives you the consensus:

```
Friends, Romans, countrymen, lend me your ears;
```

If you've read the play, you know this as the beginning of Antony's famous speech from Act III, Scene 2. If not, you might still recognize it as "real" English—it has syntax and meaning. And this is just how genome assembly works. Sophisticated computer programs use pattern recognition to look for overlaps and assemble clusters of sequences, commonly referred to as "contigs" (for "contiguous assemblies," I suppose). Smaller contigs are joined together to make larger ones, usually using other information about how they might fit together. These are generally referred to as "scaffolds." Ultimately, all of these are put together into the whole genome sequence.

However, it's not necessarily quite that simple. All but the simplest genomes (viruses and bacteria, for example) are riddled with pieces of DNA sequence that look like one another, the dreaded "repetitive elements." In our play analogy, you can think of them as common words that appear multiple times. These can play havoc with the assembly process. Consider this, one of my favorite passages of Shakespearean dialogue, from Act IV, Scene 2:

> CASSIUS: Stand, ho!
> BRUTUS: Stand, ho! Speak the word along.
> FIRST SOLDIER: Stand!
> SECOND SOLDIER: Stand!
> THIRD SOLDIER: Stand!

Imagine you were trying to reassemble the play's script from many tiny pieces, and you came across a fragment, "Stand." Where does it go? Did Cassius speak it, or does it belong to Brutus?

Even worse, what if there was an ambiguity in the last character? Perhaps it was blurred due to water damage, so you knew only that it was "Stand," followed by *something*. Is that something a comma, an exclamation mark (both of which would fit in the passage above), or something else (a letter *s*, for example)? This kind of ambiguous character happens all the time in DNA sequencing, and the result in our example above is that now our piece of text could go in one of five places, spoken by no fewer than five different characters.

We can fix this, at least some of the time, by asking for longer pieces of sequence (or, getting back to the jigsaw analogy, bigger pieces). If we have a longer text fragment like "Stand, ho! Speak the," then we know *exactly* where it's supposed to go, even though each of the four words individually occurs more than once in the play, and even if there are a few ambiguities in it. This is how we get around really short repeats in DNA sequence—by using sequence reads long enough to either encompass them entirely, or to make a unique sequence. But for long words or phrases that appear more than once, sometimes we just can't find the right location, no matter what we do.

Except—there is one more commonly used trick, called "paired-end" or "mate pair" sequencing (these are slightly different things, but for our purpose it doesn't really matter). This takes advantage of deriving DNA sequences from both ends of a much larger fragment, and using the location information from one to help position the other. Here's how it works.

The phrase "Caesar's house" is spoken twice, once by a servant in Act III, Scene 2, and again by Antony in Act IV, Scene 1. Now, imagine we have a fragment of the play of, say, half a page or so in length, and we've found "Caesar's house" at one end of it. If we look at the other end, we might find the rather unusual phrase "slanderous loads." The only place this occurs is in Act IV, almost exactly half a page away from one of the instances of "Caesar's house."

Now, we can be reasonably certain that we've properly "mapped" the first phrase onto the play, and that it's the one Antony speaks. The key here is that *we don't need to know all the text in between*—in other words, we don't need the whole half page of text to help in mapping the ambiguous end. We just need to know that the two ends are about a half a page apart. This method of using paired sequences from the ends of large DNA fragments to map non-unique sequences is very useful in assembling unknown genomes.

Genome assembly is much more complex and fraught with problems than I've led you to believe, but that's the basic idea. I'll just leave you with Octavius's closing words:

So call the field to rest; and let's away,
To part the glories of this happy day.

RICHARD F. WINTLE is a molecular biologist and geneticist. He obtained his Ph.D. from the University of Toronto, studying the human genome, and is currently the assistant director of the Centre for Applied Genomics at the Hospital for Sick Children in Toronto. When not thinking about genomes, he blogs at *Occam's Typewriter*: http://occamstypewriter.org/wintle/.

48

THE INTELLIGENT HOMOSEXUAL'S GUIDE TO NATURAL SELECTION

JEREMY YODER

June is Pride Month in the United States, and in communities across the country, lesbian, gay, bisexual, and transgendered Americans celebrate with carnivals, parades, and marches. Pride is a rebuke to the shame and marginalization many LGBT people face growing up, and a celebration of the freedoms we've won since the days when our sexual orientations were considered psychological diseases and grounds for harassment and arrest. It's also a chance to acknowledge how far we still have to go, and to organize our efforts for a better future.

And, of course, it's a great big party.

As an evolutionary biologist, I suspect I have a perspective on the life and history of sexual minorities that many of my fellow partiers don't. In spite of the progress that LGBT folks have made toward legal equality, there's a popular perception that we can never really achieve biological equality. This is because same-sex sexual activity is inherently not reproductive sex. To put it baldly, natural selection should be against men who want to have sex with other men, because we aren't interested in the kind of sex that makes babies. An oft-cited estimate from 1981 is that gay men have about 80 percent fewer children than straight men.

Focusing on the selective benefit or detriment associated with particular human traits and behaviors gets my scientific dander up, because it's so easy for the discussion to slip from what is "selectively beneficial" to what is "right." A superficial understanding of what natural selection favors or doesn't favor is a horrible standard for making moral judgments. A man could leave behind a lot of children by being a thief, a rapist, and a murderer, but only a sociopath would consider that such behavior was justified by high reproductive fitness.

And yet, as an evolutionary biologist, I have to admit that my sexual orientation is a puzzle.

There's reasonably good evidence for a genetic basis to human sexual orientation—although the search for a specific "gay gene" has had mixed results. Gene variants, or alleles, associated with an 80 percent decrease in reproductive fitness should be naturally selected out of the population pretty quickly. So why aren't all humans heterosexual?

Straight people are in the overwhelming majority, but gay men, lesbians, bisexuals, and transgendered people account for a nontrivial minority. The most recent survey I'm aware of found 7 percent of women and 8 percent of men in the United States identify as L, G, B, or T. We don't have remotely comparable historical data, but mention of same-sex sexuality goes back to the dawn of recorded history. If natural selection is homophobic, it's not particularly good at it.

A quick disclaimer: I'm going to consider how same-sex attraction might persist in human populations in the face of its apparent selective disadvantages. In the absence of direct data—such as systematic measures of the total evolutionary fitness of gay men or lesbians in specific societal contexts—it's easy to make up stories about natural selection, but much harder to determine which stories reflect reality. I'll try to delineate which stories fit with what we know about how selection works, and with the little data we do have, but that's the best I can do. Evolution is complicated, and human evolution doubly so.

Natural Selection Isn't All-Powerful

Natural selection causes traits associated with having fewer children to become less common over time. But natural selection is not the only evolutionary process at work in natural populations. Mutation introduces new alleles even as natural selection removes them. Furthermore, the effects of random chance in small populations creates an effect called genetic drift, which can interfere with the expected operation of natural selection.

Evolutionary biology has developed an excellent understanding of how mutation, selection, and drift interact over time to shape the genetic diversity of populations. That understanding allows us to do some back-of-the-envelope calculations to see how selection might operate on a gene associated with same-sex attraction. In setting this up, I'm following the lead of the evolutionary biologist Joan Roughgarden, who makes a similar point in her book *Evolution's Rainbow.* Brace yourself for some math.

In an idealized population of infinite size, the balance between natural selection's effect of removing disadvantageous alleles, and mutation's effect of spontaneously re-creating them, means that the equilibrium frequency of the allele in the population should be about equal to the square root of the ratio between the mutation rate and the selective cost associated with carrying two copies of the disadvantageous allele. Simply put, at higher mutation rates, a disadvantageous allele can maintain higher frequency; but if selection against the allele is stronger, the equilibrium frequency will be lower.

A single base pair of human DNA has a chance of mutating equal to about 1 in 100,000,000 every generation. Since there may be thousands of base pairs in a single gene, the probability of a mutation occurring somewhere in the gene is more like 1 in 100,000. If we assume that it takes two copies of our hypothetical "gay allele" to make a person attracted to members of the same sex, and about 5 percent of people are attracted to members of the same sex, then mutation alone could balance a selective cost to being gay of 0.0002. That is to say, mutation-selection balance alone could explain the frequency of LGBT folks in the population if those attracted to the same sex had, on average, 0.9998 children for every child born to the average straight parent. That's pretty weak selection.

This is where genetic drift enters the picture. Most natural populations don't behave anything like the mathematical ideal assumed for the calculations in the preceding paragraph, because most natural populations are not infinite in size. In finite populations, randomness—"mere bad luck," in the words of pioneering biologist J.B.S. Haldane—can prevent selection from operating efficiently. Smaller populations are more prone to genetic drift—the relevant number is not necessarily the number of individuals in the population, but the effective population size.

The classical estimate of the human effective population size is about 10,000, and more recent estimates have come up with even smaller numbers. This is because our population's expansion to billions is a very recent phenomenon by evolutionary standards, and may reflect the fact that for much of our history we lived in smaller, isolated populations. For populations in this size range, selection may not operate efficiently.

Setting up an experiment that would take into account the effects of drift, mutation, and selection acting together on the human population is impossible in both practical and ethical terms. That leaves biologists two ways to approach the question of how a particular disadvantageous allele can persist in the human population: intensive study of the population's genetics, and mathematical or computer-based modeling. Lacking easy access to massive amounts of human population genetic data, I've built a computer model.

The model is a script for the excellent open-source programming language R. In a nutshell, it simulates the evolution of a population of critters that may have two copies, one copy, or no copies of a deleterious allele. Critters with two copies have their chances of reproducing reduced by a set amount, which is the selective cost of carrying two copies of the allele; critters with one or no copies experience no such cost.

Every generation, the critters who survive to mate pair with randomly selected partners to form offspring, who then replace their parents to start the cycle all over again. This randomized mating in a finite population allows for drift to occur. Everyone who survives to mate has an equal chance to mate, but some are randomly paired more than once, and some miss out. Mutation occurs at the moment of reproduction, when the alleles passed on from parents to their offspring have a small chance of changing to the deleterious form, or back to the harmless form.

I set up the simulation with a population size of 10,000, in which about 5 percent of the critters carry two copies of the deleterious allele. I set the cost of carrying two copies of the allele to 20 percent, and the probability of mutation each generation to 1 in 100,000.

Here's what happens to the percentage of critters carrying two copies of the deleterious allele over fifty generations of sim-evolution. The solid line in the following graph is the percentage of critters in the population carrying two copies of the deleterious allele; the dotted line marks the starting percentage, for reference.

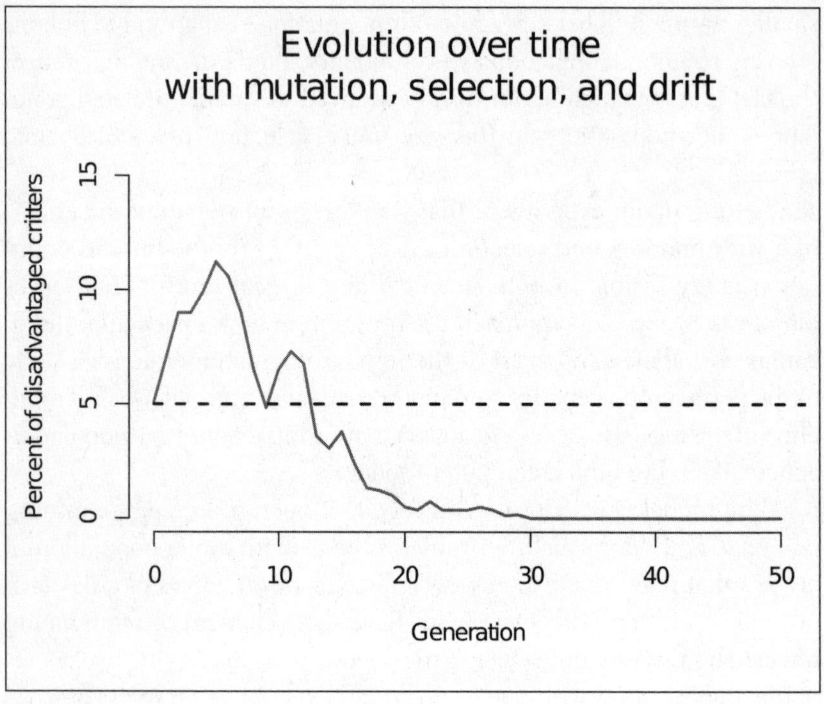

(Created by Jeremy Yoder)

You can see that selection wins out, as we'd expect, but it doesn't do so immediately. In fact, there are periods where the percentage of critters with two disadvantageous alleles increases. That's the randomness of drift in action.

To get a sense of how this drifting randomness plays out in general, we need to run the simulation many times and see what tends to happen. This is like flipping a coin over and over to see whether it really does land heads side up 50 percent of the time. The second figure shows ten replicate simulations, with graphs like the one above superimposed on one another for comparison. The third figure shows a hundred simulations.

So drift and mutation complicate things; sometimes, rarely, the disadvantageous allele persists for fifty generations. However, we can say from these simulations that selection removes the allele far more often than not. These simulations assume selection quite a bit weaker than the widely cited cost of same-sex sexuality—20 percent lower fitness instead of 80 percent—so you would expect selection to be even more effective against an allele that makes men gay and women lesbian.

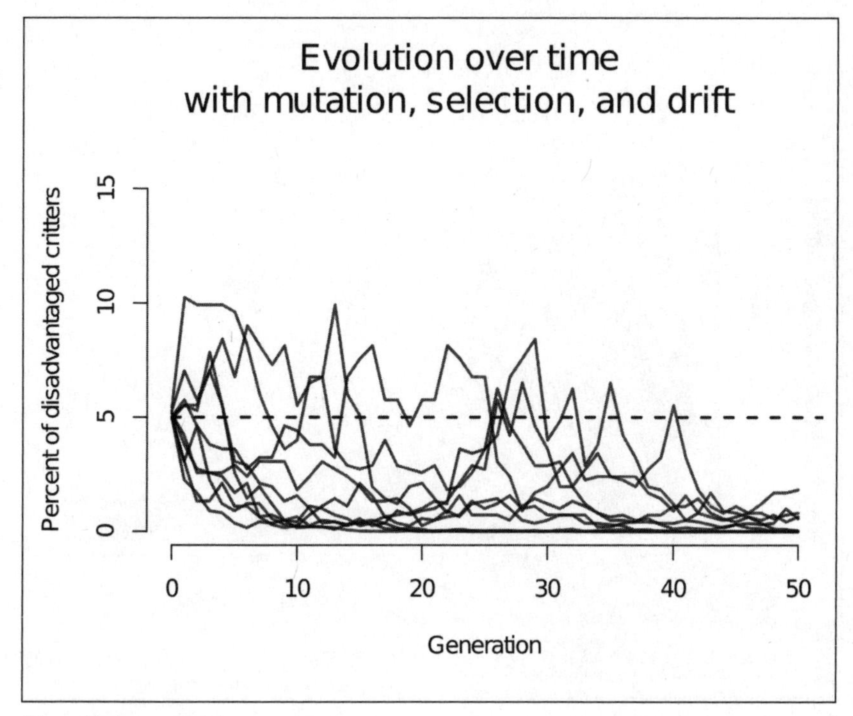

(Created by Jeremy Yoder)

However, remember that same-sex sexuality has been present in human populations for considerably longer than fifty generations. That suggests my simulations don't accurately reflect the real-world evolution of human sexual minorities—not because I've simulated weaker selection, but because I've simulated selection that works too well.

It could be that I've modeled the genetics incorrectly, by assuming that there's only a single gene involved in determining sexual orientation. However, as the evolutionary biologist Douglas Futuyma has pointed out in his review of Roughgarden's book for the journal *Evolution*, if multiple genes are involved, they would "share" the selective costs associated with same-sex attraction. Selection would then have proportionally less power to remove alleles for same-sex sexuality in the face of drift and mutation. Given that recent genome studies have not clearly identified a single gene region associated with sexual orientation, it seems likely that multiple genes are indeed involved.

The other possibility is that selection against same-sex sexuality is not as strong as I've made it in my simulations. How is that possible? Well,

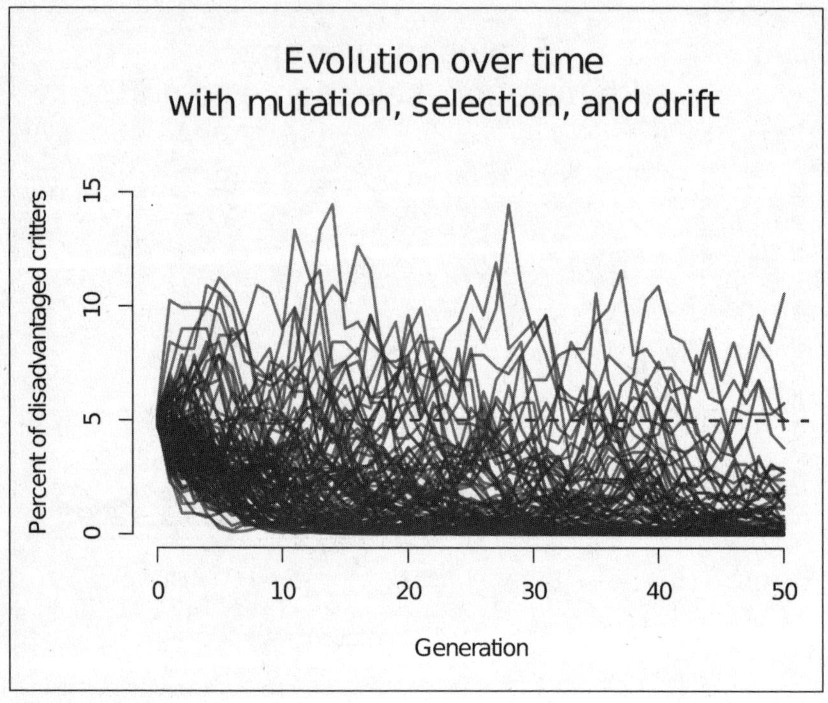

(Created by Jeremy Yoder)

simply put, the lives of gay men, lesbians, and transgendered people in Western societies over the last few decades might not be much like the lives we would live in other times and other places. And that could make all the difference.

Context Matters

In the United States, the most well-known LGBT life story goes something like this: one discovers same-sex attraction in adolescence and comes out of the closet before having much sexual interaction with members of the opposite sex. For committed same-sex partners, having biological children is possible via surrogacy or sperm donation, but it's complicated by the lack of legal recognition and protection for the couple and for children they choose to have. All these factors tend to reduce the number of biological children gays, lesbians, and trans folks have—but they're also all phenomena of our current historical and political moment. In different social contexts, LGBT fitness could very well be higher.

Before the gay rights movement, social expectations probably led many people who today would identify as gay or lesbian to enter into straight marriages and raise families. This kind of social pressure may explain why gay and lesbian couples are more likely to be raising children if they live in the conservative southern United States—not because adoption is more common in that region, but because in the South, gay men and lesbians are more likely to have heterosexual relationships, and children, before they come out.

On the other side of that coin, it's possible to imagine that in societies in which same-sex relationships receive the same legal recognition as straight marriages, gay men and lesbians might eventually have biological children at substantially higher rates than they do today. So if oppression might reduce the fitness cost to being gay, equality probably could, too.

Many non-Western societies, too, have accepted social roles for a "third gender" that can encompass identities approximating Western gays, lesbians, bisexuals, and transgendered folks. The roles and behaviors of third-gendered individuals vary considerably, but in some cases they may have biological children even if their primary relationships are with members of the same sex. As Christopher Ryan and Cacilda Jethá describe extensively in their book *Sex at Dawn*, pre-agricultural human societies may have been highly polygamous by modern standards, with children raised communally. In that context, Ryan and Jethá propose, same-sex sexual interactions may have provided social capital that could help to support children produced in relatively infrequent heterosexual couplings.

This is consistent with the picture Joan Roughgarden paints in *Evolution's Rainbow*. Roughgarden's book describes widespread same-sex activity that usually fosters social relationships in support of reproductive sex rather than instead of it. To pick just one example out of dozens, some male bluegill sunfish look and behave like females, and "court" more masculine males—then help attract a female and share in the resulting three-way reproductive opportunity by fertilizing some of her eggs. The wide range of nonreproductive sexual behavior in the broader animal kingdom suggests that human sexual minorities are just one manifestation of a phenomenon that could date back to the origins of sex itself.

Reproduction by Proxy

It has also been suggested that LGBT folks might make up for a lack of biological children by boosting the reproductive success of their close relatives. My brother shares half of my genetic material, so if I help him and his future wife support more children than they would have otherwise, those nephews and nieces "count" toward the children I'm not making myself.

This kind of indirect fitness might offset a lack of direct reproduction, but I doubt that it can cancel it out. Genetically, a nephew "counts" toward my fitness about half as much as a son does. Therefore, if I would otherwise have two children on my own, I have to help my brother to have four additional children to make up for them. I'd like to think I'll make a good uncle, but I won't be that good. And, in fact, one survey of gay men has found that they aren't significantly more generous toward their nephews and nieces than straight men are. (One objection you might make to this study is that it addresses our current societal context, not that in which humans originally evolved.)

A subtle twist on the indirect fitness idea that I haven't seen in the scientific literature could be to consider the fitness not of gay men and lesbians, but that of their mothers. In order for my mother to have as many grandchildren as she would with two straight sons, I need only help my brother and his wife support two additional children. That seems more achievable, and might work out in the context of the polyamorous, mutually helping tribes described in *Sex at Dawn*. It's also consistent with the observation that men are more likely to be gay if they have older brothers—a woman who has already had several straight sons might, conceivably, have more surviving grandchildren by giving them a helpful gay uncle.

Finally, there is some evidence that genes associated with same-sex attraction in men might provide a fitness boost in women. Since any given gene has about a 50 percent chance of ending up in one sex or the other, a gene that makes men more likely to be gay but makes women more fertile might, on average, have no selective advantage or disadvantage. A 2004 study found that women related to gay men have more children, which supports this scenario. The biologists Sergey Gavrilets and William Rice used a mathematical model of selection on same-sex sexuality to consider this hypothesis in a 2006 paper, and found that a female fertility boost could indeed allow male same-sex sexuality to persist.

In the end, however, this is mostly storytelling—lots of possibilities, but much less hard data. What we need to test many of these ideas is detailed records of the total reproductive fitness of sexual minorities in specific social contexts—especially societies approximating the ones formed by our earliest human ancestors. The best we can say without this is that many societal contexts could have made the apparent fitness cost to same-sex attraction smaller than it appears at first glance.

So where does all of this leave the evolutionarily aware gay man, lesbian, or transgendered person? Figuring out the exact nature of our tenuous relationship with natural selection doesn't tell us much about our moral stature, our value to society, or the best way to live our lives. It does, however, offer to answer the question that evolutionary biology can potentially answer for all human beings, regardless of orientation, gender, or race: how did we come to be what we are?

The best answer we have so far is complicated. It may be that we're children of history and chance, not a clear-cut adaptive path. But easy lives and clear-cut answers aren't, I think, what we celebrate in the history of our LGBT forerunners, or remember at Pride rallies. If we queer folk live our lives in the tail of a probability distribution, the good news is that the company here is pretty good.

Note: The computer simulation described in this article is freely available online at www.jeremybyoder.com/documents/MSDsim.R. To run the simulation, you'll also need a copy of the programming language R, which is available for free at www.r-project.org and runs on Windows, Mac, and Linux operating systems.

REFERENCES

Bailey, Nathan W., and Marlene Zuk. "Same-Sex Sexual Behavior and Evolution." *Trends in Evolution and Ecology* 24, no. 8 (2009): 439–46. doi:10.1016/j.tree.2009.03.014.

Bogaert, Anthony F. "Biological Versus Nonbiological Older Brothers and Men's Sexual Orientation." *PNAS* 103, no. 28 (2006): 10771–74. doi:10.1073/pnas.0511152103.

Camperio-Ciani, Andrea, Francesca Corna, and Claudio Capiluppi. "Evidence for Maternally Inherited Factors Favouring Male Homosexuality and Promoting Female Fecundity." *Proceedings of the Royal Society B* 271, no. 1554 (2004): 2217–21. doi:10.1098/rspb.2004.2872.

Eller, Elise, John Hawks, and John H. Relethford. "Local Extinction and Recolonization, Species Effective Population Size, and Modern Human Origins." *Human Biology* 76, no. 5: 689–709. doi:10.1353/hub.2005.0006.

Futuyma, Douglas J. "Celebrating Diversity in Sexuality and Gender." *Evolution* 59, no. 5 (2005): 1156–59. doi:10.1111/j.0014-3820.2005.tb01052.x.

Gavrilets, Sergey, and William R. Rice. "Genetic Models of Homosexuality: Generating Testable Predictions." *Proceedings of the Royal Society B* 273 (2006): 3031–38. doi:10.1098/rspb.2006.3684.

Haldane, J.B.S. "A Mathematical Theory of Natural and Artificial Selection, Part V: Selection and Mutation." *Mathematical Proceedings of the Cambridge Philosophical Society* 23, no. 7 (1927): 838–44. doi:10.1017/S0305004100015644.

Hamer, Dean H., Stella Hu, Victoria L. Magnuson, Nan Hu, and Angela M. L. Pattatucci. "A Linkage Between DNA Markers on the X Chromosome and Male Sexual Orientation." *Science* 261, no. 5119 (1993): 321–27. doi:10.1126/science.8332896.

Herbenick, Debby, Michael Reece, Vanessa Schick, Stephanie A. Sanders, Brian Dodge, and J. Dennis Fortenberry. "Sexual Behavior in the United States: Results from a National Probability Sample of Men and Women Ages 14–94." *Journal of Sexual Medicine* 7, no. s5 (2010): 255–65. doi:10.1111/j.1743-6109.2010.02012.x.

Pillard, Richard C., and J. Michael Bailey. "Human Sexual Orientation Has a Heritable Component." *Human Biology* 70, no. 2 (1998): 347–65. PMID: 9549243.

Rahman, Qazi, and Matthew S. Hull. "An Empirical Test of the Kin Selection Hypothesis for Male Homosexuality." *Archives of Sexual Behavior* 34, no. 4 (2005): 461–67. doi:10.1007/s10508-005-4345-6.

Ramagopalan, Sreeram V., David A. Dyment, Lahiru Handunnetthi, George P. Rice, and George C. Ebers. "A Genome-Wide Scan of Male Sexual Orientation." *Journal of Human Genetics* 55, no. 2 (2010): 131–32. doi:10.1038/jhg.2009.135.

Rice, George, Carol Anderson, Neil Risch, and George Ebers. "Male Homosexuality: Absence of Linkage to Microsatellite Markers at Xq28." *Science* 284, no. 5414 (1999): 665–67. doi:10.1126/science.284.5414.665.

Roach, Jared C., Gustavo Glusman, et al. "Analysis of Genetic Inheritance in a Family Quartet by Whole-Genome Sequencing." *Science* 328, no. 5978 (2010): 636–39. doi:10.1126/science.1186802.

Roughgarden, Joan. *Evolution's Rainbow: Diversity, Gender, and Sexuality in Nature and People*. Berkeley: University of California Press, 2005.

Ryan, Christopher and Cacilda Jethá. *Sex at Dawn: The Prehistoric Origins of Modern Sexuality*. New York: Harper, 2010.

Takahata, Naoyuki. "Allelic Genealogy and Human Evolution." *Molecular Biology and Evolution* 10, no. 1 (1993): 2–22. PMID: 8450756.

Tenesa, Albert, Pau Navarro, Ben J. Hayes, David L. Duffy, Geraldine M. Clarke, Mike E. Goddard, and Peter M. Visscher. "Recent Human Effective Population Size Estimated from Linkage Disequilibrium." *Genome Research* 17, no. 4 (2007): 520–26. doi:10.1101/gr.6023607.

Xu, Lin, Hong Chen, Xiaohua Hu, Rongmei Zhang, Ze Zhang, and Z. W. Luo. "Average Gene Length Is Highly Conserved in Prokaryotes and Eukaryotes and Diverges Only Between the Two Kingdoms." *Molecular Biology and Evolution* 23, no. 6 (2006): 1107–8. doi:10.1093/molbev/msk019.

JEREMY YODER is a postdoctoral associate studying plant population genetics at the University of Minnesota in Saint Paul. You can find more of his writing about evolution and ecology on his blog *Denim and Tweed*, at www.denimandtweed.com.

49

THE RENAISSANCE MAN

ED YONG

Erez Lieberman Aiden is a talkative, witty fellow, who will bend your ear on any number of intellectual topics. Just don't ask him what he does. "This is actually the most difficult question that I run into on a regular basis," he says. "I really don't have anything for that."

Aiden is a scientist, but while most of his peers stay within a specific field—say, neuroscience or genetics—Aiden crosses them with almost casual abandon. His research has taken him across molecular biology, linguistics, physics, engineering, and mathematics. He was the man behind 2010's "culturomics" study, in which he looked at the evolution of human culture through the lens of 4 percent of all the books ever published. Before that, he solved the three-dimensional structure of the human genome, studied the mathematics of verbs, and invented an insole called the iShoe that can diagnose balance problems in elderly people. "I guess I just view myself as a scientist," he says.

His approach stands in stark contrast to the standard scientific career: find an area of interest and become increasingly knowledgeable about it. Instead of branching out from a central specialty, Aiden is interested in problems that cross the boundaries of different disciplines. His approach is nomadic. He moves about, searching for ideas that will pique his curiosity, extend his horizons, and, he hopes, make a big impact.

"I don't view myself as a practitioner of a particular skill or method," he says. "I'm constantly looking at what's the most interesting problem that I could possibly work on. I really try to figure out what sort of scientist I need to be in order to solve the problem I'm interested in solving."

It's a philosophy that has paid rich dividends. In 2010, at thirty years of age, Aiden got a joint Ph.D. in applied math and bioengineering at MIT and Harvard and won the prestigious $30,000 Lemelson-MIT Student Prize, awarded to people who show "exceptional innovation and a portfolio of inventiveness." Now a Harvard fellow and visiting faculty at Google, he has seven publications to his name, six of which appeared in the world's top two journals—*Nature* and *Science*. His friend and colleague Jean-Baptiste Michel says, "He's truly one of a kind. I just wonder about what discipline he will get a Nobel Prize in!"

When I meet Aiden at Harvard, he is dressed casually in a jersey, chinos, and trainers. He talks quickly but eloquently, at once relaxed and in deep concentration. The door to his office opens into a room that feels more like a lounge. In place of benches and stools, there is a comfortable sofa, armchairs, several computers, and a big TV. Aside from a pile of snacks, the space is notably spartan. There are no photos on the walls. Three rows of shelves are largely empty. The desks are unburdened. The room, like the man himself, is uncluttered by the past.

Aiden naturally gravitates to problems that he knows little about. "The reason is that most projects fail," he says. "If the project you know a lot about fails, you haven't gained anything. If a project you know relatively little about fails, you potentially have a bunch of new and better ideas." And Aiden has a habit of using his failures as springboards for success.

In 2005, Aiden was fascinated by the way we make antibodies. Antibodies are all very similar, but their tips—the bits that recognize invaders—are extremely variable. These are created through a genetic pick-and-mix: genes from three groups, each with many different members, are united in one of 100 million different combinations. This vast range of permutations provides the variety we need to counter a legion of threats from bacteria, viruses, parasites, tumor cells, and more. "The immune system constantly creates genes on the fly that are specific to the things that show up in the body. It's amazing," says Aiden. His goal was ambitious but simple: catalog these genes and sequence the human immune system.

He failed. "The problem is that all the genes are very, very similar," he says. Sequencing genes isn't like reading text from start to finish. It's more like looking at isolated fragments of sentences and trying to join them into the original narrative. If the sentences all contain roughly the same words, that task becomes very difficult. "At a certain point, we realized that the data just wasn't good enough. That was a disaster—it was eighty-five percent of my time for eighteen months. That was an epic failure."

But it was not a wasted opportunity. In 2007, Aiden's interest in antibodies took him to an immunology conference, where he accidentally went into the wrong talk. There, Aiden found the inspiration that would lead him to solve the three-dimensional structure of the human genome.

The speaker, Amy L. Kenter, was discussing the physical distances between our genes. Each of our cells has the unenviable task of folding a six-foot-long stretch of DNA into a chamber about a million times shorter in diameter. The cells do it by folding DNA into complex shapes, a feat of origami that often turns distant genes into close neighbors. Aiden learned that these distances were very hard to calculate. People would spend up to six months figuring out the distance between two sites. "It prompted a knee-jerk response," he says. "I was completely convinced that what they were doing could be done better and faster."

To speed up the process, Aiden invented a technique called Hi-C that simultaneously identifies neighboring sites across the entire genome. First, he embalms the genome with formaldehyde. The chemical creates physical bridges between different pieces of DNA that lie next to one another, freezing the genome in all its twists and turns. Special enzymes shred the DNA, and the fragments are isolated, sequenced, and mapped onto the reference copy of the human genome. The result is a massive library of interacting DNA—a genetic social network. Aiden could then work out how the genome must have folded to accommodate these interactions.

He found something odd. Polymers—long-chain molecules, such as DNA—tend to fold in predictable ways. They ought to form densely packed and knotted bundles called "equilibrium globules"—think of a plate of cooked noodles, or headphones that have been left in a pocket for too long. But the Hi-C results weren't compatible with this shape; they suggested that the genome was doing something different. At first, Aiden thought his technique had failed. He started reading voraciously, absorbing everything he could find about polymer physics. And every source led to the same conclusion: his results seemed to violate established physical principles.

His breakthrough came in the dead of night. He discovered a paper by a physicist named Alexander Grosberg, who described a shape called a fractal globule. It, too, is a densely packed bundle, but unlike the equilibrium globule, it doesn't have a single knot. The strands may loop and twist, but they never cross and tangle. Aiden likens it to uncooked noodles—you can pull out one strand without disrupting the rest.

The fractal globule was first described by an Italian mathematician, Giuseppe Peano, in 1890, but it was completely theoretical. It took almost a century for Grosberg to suggest (in 1988) that a real polymer might fold into a fractal globule if the conditions were right. In 2009, Aiden proved him right. "I read [Grosberg's paper] and I immediately thought: this solves it!" The fractal globule makes perfect sense as the shape of a genome. With no tangles, any stretch of DNA can be easily exposed so that its information can be transcribed and used. "That was one of the most exciting moments of my life intellectually," says Aiden.

As far as anyone knew, the fractal globule was a hypothetical shape that existed only in Peano's imagination. Aiden showed that it exists inside every human that has ever walked the earth. "One is not reasonably entitled to expect that one's data is going to happen to be consistent with some age-old dead hypothesis that ends up being much more beautiful than the dominant idea," he says with a wry smile. "That's just pennies from heaven."

All of this came from the failed project on antibodies. Aiden's cutting room floor is filled with equally ambitious dead projects on the evolution of Chinese iconography and network analyses of people suing each other. In most cases, they simply became too boring to continue, but rare cases like the 3-D genome paid off. "The best types of problems are those that seem harder at first than when you think about it. If you have ten such projects and one of them works, you're good because lots of people think it's astronomically unlikely the project would have worked and they don't know you've tried ten of them," he says.

"The failures very naturally lead to new successes and opportunities. That's why it's great to get a couple of failures under your belt in a new area. The immunology project was the first big genomics project that I really sunk my teeth into, and all of the tools that I picked up during that failure turned out to be very useful in the 3-D genome sequencing."

The 3-D genome project epitomizes many of the themes that run through Aiden's diverse oeuvre. He has a great belief in the power of technological advances. "Much of contemporary science is really the length and shadow of the technology we apply," he says. By inventing the Hi-C technique, he could ask questions about the genome that simply weren't possible to answer before. "I'm always on the lookout for new methods that I think will open up whole new domains." In particular, he likes to accumulate large sets of data without any preconceptions. "For me, seeing is believing. I rarely have any hypotheses when I start looking at a data set. I'm just trying to see what features jump out at me."

Aiden's mind-set runs in the family. His son, Gabriel Galileo, is just one year old and shares his father's consuming curiosity. "He's figuring out the fundamental things that challenged humans. Billions of years to work out how to balance on two feet and he's like, 'Well, that's Thursday.'"

As a child himself, Aiden learned the value of being curious and well-rounded from his father—a technology entrepreneur named Aharon Lieberman. "I spent many days and even summer months working with him in his factory," says Aiden. "The idea that one could support oneself by making ideas a reality is one my dad always emphasizes. He gave me a lot of self-confidence. This helps, because when you suddenly change the subject in your work, all you take with you are your brains and your confidence in your own ability to figure things out."

As an undergraduate, he studied mathematics, physics, and philosophy at Princeton. "My reasoning was that I would be able to figure out the universe and make all subsequent life decisions from first principles," he says, grinning. "It was the kind of thing that makes sense to you in high school. Oh yeah, everything is going to reduce to quantum mechanics and you can work it out . . . Anyway, that was a disastrous failure." And once again, the quest to "debug this failure" led to something interesting.

"It turns out you can't work everything out from first principles, because it seems like a lot of things have happened and I didn't know anything about the universe before I was born in 1980," he says. "So I thought, I have to go and understand that stuff." He spent a year at New York's Yeshiva University studying for a master's in history. He took classes going back in time from the present day, read forward from ancient history (he can now read Aramaic), and stopped when the two streams met in the seventeenth century.

Eventually, Aiden returned to the sciences, securing a master's in applied physics at Harvard and a Ph.D. in applied math and bioengineering at Harvard and MIT. But his foray into the humanities had also left its permanent mark. His most ambitious project to date—culturomics—is very much a fusion of the so-called two cultures.

Once again, it began with a talk, this time by Steven Pinker. Pinker mentioned that while just 3 percent of English verbs are irregular (such as *be* or *do*), they are the most commonly used ones. All the ten most commonly used verbs are irregular. For Aiden, who had long been thinking about studying culture in a mathematical way, this piece of trivia was irresistible. Together with Jean-Baptiste Michel, he charted the course of irregular verbs from the ninth-century *Beowulf* to the thirteenth-century *Canterbury Tales* to the twenty-first-century Harry Potter series. They focused on 177 irregular verbs and found that the verbs "regularize" with time, with rarer verbs falling into line more quickly.

This road to conformity can be described by a very simple mathematical formula. Verbs regularize in a way that is "inversely proportional to the square root of their frequency." If one is used a hundredth (.01) as frequently as another, it will become regular ten times as fast. If it's used only a millionth (.000001) as frequently, it will regularize a thousand times as fast. Based on how frequently a verb appears, you can predict when it will yield to regularity. "Read" is unlikely to change to "readed" anytime soon, but "burnt" is rapidly being cast aside in favor of "burned."

The result was fascinating, but scouring old books was an unenviable task. "The data collection took a year and a half. It was a huge pain and a Hail Mary because we never knew whether it would work," says Aiden. "At the end of it, we said, 'We can never do this again.'" Fortunately, they never had to. As the paper went to press, Aiden went back to his Middle English texts to check his facts and realized that, in the meantime, someone else had taken them out of the library—Google.

In 2004, Google began digitizing the world's books, in an ambitious project that, as of 2011, has since scanned more than 15 million books from over forty university libraries. This online corpus represents 12 percent of all the books ever published, a massive electronic record of humanity's culture. "On some level, this was phenomenally embarrassing," says Aiden. "We realized our methods were so hopelessly obsolete. It was clear that you couldn't compete with this juggernaut of digitization."

So instead of competing, Aiden and Michel decided to join them. Their pitch was simple: they would use the words in Google's corpus to track the path of culture over time, just as paleontologists use fossils to deduce the evolution of living things. Peter Norvig, Google's director of research, was sold from the first meeting.

However, there were serious obstacles. "Midway through the project, Google gets sued by absolutely everybody," says Aiden. "That's not helping." There were also problems with the data. In some cases the scans weren't clear enough, and in others, the "metadata," such as the dates of publication, were inaccurate. This meant that words like "Internet" would seem to turn up well before such a thing was conceived.

It took a year to clean up the data, and still there were imperfections. Eventually, Aiden and Michel restricted themselves to a third of the corpus—some 5 million books in six languages. They pulled out billions of individual words and phrases ("ngrams") and tracked their frequency over time, compiling everything into a large data set that anyone can download and explore.

At the time, Aiden wrote, "Together, these furnish a great cache of bones from which to reconstruct the skeleton of a new science." He named the science "culturomics"—the quantitative study of human culture. It was envisioned as the cultural equivalent of the human genome project—a treasure chest of data to be pored over by scholars or by more casual users, via Google's popular Ngram Viewer.

Michel and Aiden revealed culturomics to the world in 2010, with a paper that offered a tasting platter of the ngrams' potential. It showed the expanding nature of the English language's lexicon and the evolving nature of its grammar. It shows "men" and "women" converging in frequency, new technologies permeating culture with quickening pace, and celebrities rising to ever-higher peaks of fame but falling from them faster. The paper even reveals the traces of suppression and censorship—"Tiananmen Square" is suspiciously absent from Chinese texts after 1989, as are Jewish artists and academics from German texts during Nazi rule in Germany.

The new approach was eye-opening, but it was inevitable that it would draw fire. "There were significant subgroups within the humanities that were up in arms," says Aiden, "because there were no humanists or historians on the paper." Such criticisms were perplexing to a man who regularly jumps from field to field. "[Qualifications] never even occurred to me as something that's relevant," he says. By contrast, when he published the 3-D genome paper, his most advanced degree was his master's in history. "No one gave a damn in the sciences!"

Other critics focused on the problems with the data, which users of the Ngram Viewer discovered for themselves. Aiden finds that frustrating. "We said in the paper that there are huge issues with the data outside the 1800 to 2000 range, but it's like if you get a TiVo or a Wii, you don't spend time reading the instructions. You just want to play with it. My hope is that people who are doing this for serious purposes eventually get the value of the tool."

Several people certainly have, and Aiden has countless examples that vindicate the project's value in his eyes. "[Alexis Madrigal] at the *Atlantic*, instead of writing a column on the nuclear age, collected a bunch of ngrams about it. These things are so clear and visual and transparent that people get that this is a way for the general public to learn a bit of history."

There have been more substantial uses, too. Wikipedia compared the quality of its articles on scientists to how famous those scientists are, as measured through ngrams. "There's a strong effect. People who are more famous have better Wikipedia articles. That's a good control. It shows that their editors have a good sense of what's important." But the analysis found something more unusual. It suggested that female scientists have consistently worse articles than their comparably famous male peers. "People talk about the fact that fifteen percent of Wikipedians are female, and that has the potential to introduce so much bias into Wikipedia itself. You could speculate about that, but now you can measure and check it."

Aiden isn't done with culturomics. He and Michel are now visiting faculty at Google ("We have access to almost all their data, so that opens up a lot of doors"). They have started a group at Harvard called the Cultural Observatory, with the aim of creating more powerful sets of data like the ones that power culturomics. And Aiden is even working on a musical version that looks at scores across time.

Once again, data quality is a big issue—musical scores are poorly annotated—but Aiden's experience in unrelated fields is yielding unexpected benefits. One of the technical challenges he solved while working on his failed immunology project turns out to be "identical," he says, to a problem with annotating scores. These are the moments that justify his nomadic career. "If we're in a room and we're talking about X, the X specialist will know more about X than I do, but I'll know more about not-X. Every once in a while, something that's not-X turns out to be very relevant."

This comes at a price: it is hard to hit the ground running in a new area, and Aiden often finds himself playing catch-up. But his broader horizons compensate for this drawback. "People have this romantic notion of inventors as people who go into caves and come out with an amazing thing that's totally novel. I think a huge amount of invention is recognizing that A and B go together really well, putting them together, and getting something better. The limiting step is *knowing* that A and B exist. And that's the big disadvantage that one has as a specialist—you gradually lose sight of the things that are around. I feel I just get to see more."

Aiden's approach harkens back to an older era for the sciences, when polymaths like Leibniz and Newton commanded respect in a variety of different fields. Such people are a rare breed in today's world, where the widening frontiers of scientific knowledge funnel scientists into narrow specialist channels. The intellectual nomads are being squeezed out.

But Aiden senses that the balance is shifting, and the connective power of the Internet plays a large part in that. "Thirty years ago, you didn't know what was going on in a different field and you did not have Google. It could take you months to figure out that an idea was a good or bad one. These days, you can get a good sense of that in a matter of minutes because information is much more accessible. That's really, really huge. It makes it much easier to move from one field to another."

The free flow of information not only makes it easier to work out which problems are available and tractable, it also makes it clear how many problems there still are—enough to fill a rich career of discipline hopping. "I had this feeling out of graduate school that everything had been done," says Aiden. "Now I think, wow, we don't know anything yet."

ED YONG is an award-winning science writer. His blog *Not Exactly Rocket Science* won the 2010 National Academies / Keck Futures Initiative Communication Award and was represented in the *Best American Science Writing 2011* anthology. His work has also appeared on the BBC and CNN, and in *The Daily Telegraph, Discover, The Economist, The Guardian, Nature, New Scientist, Slate, The Times* (London), *Wired UK*, and more. He lives in London with his wife.

50

FRAGMENTED INTIMACIES

SHARA YURKIEWICZ

His face was four inches away from mine. I tried not to blink as he shined the ophthalmoscope's light into my left eye and stared into my pupil as though it were the most interesting thing in the world. He frowned, placed his hand on my head, and used his thumb to pry my eyelid higher. He maneuvered for about forty-five more seconds while I sat stone still, and then, suddenly, his face broke into a grin. "I see it," he announced. "I definitely see it."

And then, completely awestruck, "Wow."

I was my classmate's first visualization of the optic disc.

Our ophthalmology instructor previously had shown us dozens of images of the inside of the eye, some normal, some frighteningly abnormal. "Before I say anything else," he began, "the first thing I want you to notice is how beautiful the eye is."

Indeed it was. Snaking along the back of the eye were tiny red delicate blood vessels, converging to become thicker until they crossed over the optic disc. The optic disc, a pale yellow standout among the redder hues, is unique because it is the only part of the central nervous system that we can noninvasively see.

When my classmate saw my optic disc, he saw a piece of me that no one had ever seen before. He saw my central nervous system.

Later that night, I friended him on Facebook.

Medical school is an intimate experience. When we learned to test for reflexes, I unabashedly hit my partner's forearm until it bruised. When studying for my microbiology final, I sat in the computer lab eating and complaining with a few other students, finally shuffling out with them at 1:00 a.m. When I interviewed my first patient in the hospital, my partner watched me stumble over the most basic questions—and later told my preceptor he thought I did a great job.

What has been surprising is the complete lack of continuity among these experiences. I have parents, I have a sister, I have a best friend, I have close friends in my class, I have had the significant other. These people see me as a person with a complete set of experiences. They have seen me at my highs and lows and middles, they have listened and relistened to my secrets and fears, they have offered me unconditional support.

But no matter how hard I try to paint a scene with words, they remain several degrees of separation away. They see my world through my eyes, their sense of vision stemming from my words.

My classmates have seen me on the front lines. They have seen a side of me that deals with the emotional or bizarre or just plain new. But for each experience, it is a different and sometimes unfamiliar person who shares it with me. I am used to creating memories with those closest to me, but in medical school, when the assignments are as random as the patient encounters, no such coherency exists. The thread of experiences continues, but it is fragmented. I am the only witness to the whole.

I went to an Alcoholics Anonymous meeting with two of my classmates one night as part of an assignment. We walked back slowly in the darkness, discussing, avoiding one another's eyes. In that moment, all three of us felt that it was just a bit of fortune that separated us from those whom we had met that night. We were grateful. We were empathetic. We had changed.

During anatomy, I was partnered with three different classmates. Together, we explored every cavity of our cadaver's body. We made off-color jokes. We retracted skin and guided one another's cuts. We held the lungs in our hands and marveled. We scraped the skin off the face and sawed through the skull and disengaged ourselves to do so. Six weeks later, we had changed.

This year, during a psychiatry clinic, three different classmates and I interviewed a woman with paranoid schizophrenia. Her story involved violence, abuse, homelessness, isolation, and denial. The narrator herself, by definition, was unreliable, to the point of ironically insisting that paranoia was the wrong diagnosis. It was sad. She was sad. We were sad. And afterward, we had changed.

I am changing, more than I had imagined, and I have classmates who are changing with me, sharing experiences that I can't do justice to with words when I am talking to those closest to me outside medicine. I am grateful that our paths converge, if only briefly, for those intense moments. I wonder how they perceive those moments, as time eventually blurs the details. I'm sure they have stories of their own: their individual journeys at their individual paces. But in my story, I am the only one who can put the fragmented pieces together.

Sometimes the story is lonely. Sometimes the story is exhilarating. But, ultimately, it's mine, and mine alone.

SHARA YURKIEWICZ is a medical student at Harvard University who graduated from Yale University with a B.S. in biology. She was an AAAS Mass Media Fellow and has written for various publications, including the *Los Angeles Times* and *Discover*. She is interested in medical ethics and has contributed to projects at Harvard, the Hastings Center, and the American Medical Association's ethics journal, *Virtual Mentor*. Her blog can be found at http://blogs.plos.org/thismayhurtabit/.

51

THE HUMAN LAKE

CARL ZIMMER

In 2011, I went to San Francisco to talk to a conference of scientists. These scientists are experts at assembling mountains of information—genomes, experimental results, clinical trials, and the like—and then finding something useful in them. In those mountains may be a new way to diagnose cancer, for example, or perhaps even an antibiotic drug. The invitation was an honor, but a nerve-racking one. As a science writer, I had no scientific results to offer. But we science writers have an ace in the hole.

Instead of being lashed to a lab bench for years, carrying out experiments to illuminate one particular fold in one particular protein, we get to play the field. We travel between different departments, different universities, different countries, and—most important of all—different disciplines. Sometimes we see links between different kinds of science that scientists themselves can miss. Which is why, when I arrived in San Francisco, walked up to the podium, and switched on my computer, I presented my audience with a photograph of a lake.

For the next hour, I tried to convince them that their bodies are a lot like this lake, and that appreciating this fact could help them find new ways to treat diseases ranging from obesity to heart disease to antibiotic-resistant bacterial infections.

The lake is called Linsley Pond, and it seems utterly ordinary. It is located in southern Connecticut, a short drive east of New Haven. It's about half a mile wide. It supports a typical assortment of species, including algae and bacteria, water fleas, lily pads and other aquatic plants, birds, turtles, and fishes. It does not look like a scientific treasure.

But in the history of ecology, it's one of the most significant places on Earth.

If you had gone to Linsley Pond seventy years ago, you might have seen a gentleman swimming across the lake, dumping a container of radioactive phosphorus into the water. The swimmer's name was G. Evelyn Hutchinson. Hutchinson is generally considered by ecologists to be the father of modern ecology. Before Hutchinson, ecology was, to a large extent, natural history. Naturalists would go out into the wild, catalog different species, and make a few observations. After Hutchinson, ecology became a science based on theory and mathematics, a science that asked fundamental questions about how nature works.

Lakes turned Hutchinson into a theoretician. When he began studying them, he found lakes to be like self-contained worlds, and he was fascinated by the way neighboring lakes could support different ecosystems. After he came to Yale in 1931, Hutchinson began making regular trips to Linsley Pond with his students to run experiments in order to figure out why this one lake had its particular balance of species. In effect, Hutchinson made Linsley Pond his laboratory flask.

The life in Linsley Pond is made possible by an interplanetary flow of energy. The flow starts 93 million miles away in the heart of the Sun and hurtles through space until it reaches Earth, plows through the atmosphere, and smashes into molecular traps laid out by the plants and algae in Linsley Pond. The organisms tuck away some of that energy in their cells, while some of it is released as heat.

Hutchinson and his colleagues traced the energy as it continued its flow through the lake's ecosystem, as grazing zooplankton ate the algae, as larger animals fed on the smaller ones, as they died and were, in turn, scavenged by worms and bacteria. The scientists drew diagrams to map the flow from one set of species to another. But even their most complex maps were faint shadows of the real ecosystem in Linsley Pond. There are about 200 species of zooplankton, for example, and maybe 1,000 species of algae.

Hutchinson realized that all this diversity presents a paradox. Why should each section of a food web have so many species rather than just one? Why doesn't one species outcompete all the others for that spot? Why do we have food webs instead of food chains?

Hutchinson concluded that species must slice up an ecosystem into many ecological niches. Previous generations of scientists had talked about niches before, but they had used the word pretty crudely. A "niche" might just refer to the place where a particular species lived—the altitude on a mountainside where you might find a flower growing. Hutchinson had a much more sophisticated conception of niches, one that shaped how ecologists have thought about diversity ever since.

A niche, Hutchinson proposed, was a slice of multidimensional space. One dimension might be the range of temperatures in which a species could survive. Another dimension might be the size of food particles an animal could fit in its mouth. If two species occupied different slices of ecological space, they wouldn't directly compete with each other. Thanks to this ecological space, a food web can be loaded with seemingly identical species.

Even a lake as small as Linsley Pond offers a very complicated ecological space. As you go down into the lake, the levels of phosphorus, oxygen, and other molecules change in complex ways. At every depth, in other words, you find new niches. Those niches also change through the seasons. Some species can adapt to the conditions that exist in the summer, while other specialize on other times.

Linsley Pond did not exist in an eternal cycle, however. Twenty thousand years ago, it didn't even exist. Southern Connecticut was sitting under a glacier. When the ice retreated, it left behind gouged scoops. This particular scoop filled with freshwater and became a lake. Once exposed, it didn't immediately become as it is today. It went through a process called ecological succession. There were certain species that could come into the lake quickly and take over open niches. But as they grew and reproduced, they changed the ecosystem itself. They were changing the chemistry of the lake, they were changing its transparency, they were adding to the sediment at the bottom when they died. The niches themselves changed, allowing new species to arrive in the lake and thrive.

Ecosystems tend to follow certain rules of succession. Lakes in the same region will tend to end up looking very much alike, even if they're initially colonized by different species. But there's a certain amount of luck involved, too. If the chemistry of the underlying rock and soil is different, different ecosystems will emerge. On remote islands, the mix of chance and fate is particularly striking. No land mammals ever arrived on the islands of Hawaii before humans, rats, and pigs. So there were no big predators there. On the other hand, even without mammals, Hawaii did give rise to big plant-grazers. Instead of cows, giant flightless geese filled that niche.

As ecosystems develop, they also become more resilient. They gain the ability to withstand shocks. A disease outbreak doesn't bring a resilient ecosystem crashing down; it holds together even if one species becomes extinct. The resilience of ecosystems is not infinite, however. If you push an ecosystem hard enough, it can flip to a new state.

Hutchinson and his colleagues were able to watch this kind of change in Connecticut lakes thanks to the comings and goings of a fish called the alewife. Before Europeans arrived, alewives were common to many Connecticut lakes, swimming into them each year to spawn. But dams and other changes to the land cut the fish off from many lakes, which became alewife-free. Later, as farming declined in New England, some dams came down, and fish started returning.

Hutchinson and his students realized that the return of alewives was a natural experiment. They measured the size of algae-grazing zooplankton before the arrival of alewives in a Connecticut lake, and then afterward. They saw a striking shift in the populations of the zooplankton. The alewives had wiped out the large ones, leaving the small ones to thrive. It was surprising that a species could arrive in an ecosystem and quickly exert such a powerful top-down effect on it. And the effects extended beyond the zooplankton to change the sizes of algae and food particles that were getting eaten most.

We humans have also been changing Connecticut's lakes. Linsley Pond is now lined with houses, for example. With the growing presence of humans, so comes greater erosion into the lake, and with that erosion more nutrients like phosphorus. As a result, algae populations sometimes explode, clouding the once-clear waters.

Here, then, is one way of looking at life, at how the natural world works. I would wager most of it was fairly new to the people I spoke to at the conference in San Francisco. They came from a different tradition—one that was coming into being exactly at the same time Hutchinson was swimming around Linsley Pond. Hundreds of miles away, in Tennessee, a German refugee named Max Delbrück was trying to get down to the essence of life. He looked not in a lake, but in a petri dish.

Delbrück came to the study of life from physics. He had studied with the great architects of quantum physics in the 1920s, and he became fascinated with living things from a physical point of view. How is it that they manage to retain so much order—not just through their lifetimes, but across generations? There was some understanding at the time that genes made all this possible. But nobody really knew what genes were. Many treated them as a mathematical abstraction rather than as physical things.

A gene, Delbrück speculated, is some kind of polymer arising from some kind of repeating atomic structure. It was too small for Delbrück to handle, so he wanted to find an indirect way to study it. He was certainly not going to go to Linsley Pond and work with Hutchinson. To Delbrück, that was just chaos. He wanted to study genes in a single organism. He tried flies, but they were too big and messy for him. So he shrank his focus down even further, to viruses.

Delbrück couldn't see viruses when he began his research in the late 1930s. Microscopes weren't yet powerful enough. Instead, Delbrück developed a brilliant system of infecting *E. coli* with viruses. He would be able to measure the rate at which these viruses were infecting their hosts, just by looking at the little pools of dead bacteria that grew over the course of hours.

This method allowed Delbrück to get clues to how viruses reproduced, and even how their genes mutated. By studying viruses and *E. coli*, he helped to build modern molecular biology. Delbrück's influence radiated out from those initial experiments thanks to a summer course he ran at Cold Spring Harbor Laboratory on Long Island in New York. James Watson learned genetics by studying Delbrück's phages, for example—and then, soon after, discovered the structure of DNA.

Delbrück received a Nobel Prize in 1969 for his work, and when he gave his Nobel lecture, he felt fairly satisfied. "We may say in plain words, 'This riddle of life has been solved,'" he declared.

The truth was that molecular biologists did not have just a few details left to sweep up. It would not be until 2001 that the human genome would finally be sequenced, and today, a decade later, those 3.5 billion base pairs still stubbornly hold on to many mysteries. But even if we understood the function of every gene in the human genome, we would still not understand a great deal about how the human body works. That's because the human body is not merely an oversize virus—simply a bundle of genes in a protein shell. We are, each of us, also a lake.

It is hardly news that the human body is an ecosystem. Over three centuries have passed since Antoni van Leeuwenhoek scraped off some of the scum from his teeth, mixed it in some water, put it under a microscope, and discovered "wee animalcules" swimming around in it. Ever since, scientists have tried to study the microbes that live in us and on us. But it hasn't been easy. Most species of microbes that live in our body don't do well outside of it.

In the late 1800s the German pediatrician Theodor Escherich wanted to find a better way to treat infants who were dying in droves from dysentery. He recognized that the bacteria that were killing the babies were probably living alongside bacteria that weren't harming them at all. He had to figure out the difference between them. Escherich got the stool out of healthy babies' diapers and cultured their microbes. There was one bacterium that leaped forward, as if to say, "Me, me, me, look at me!" It now bears his name, *Escherichia coli*.

E. coli became such a publicity hound thanks to its ability to grow nicely in the oxygen in Escherich's lab and to eat anything he gave it. He fed it blood, he fed it bread, he fed it potatoes. All were delicious to *E. coli*. Thanks to this eagerness, scientists (Delbrück included) began to use *E. coli* to run many of their experiments on the fundamental nature of life. And as a result, *E. coli* is arguably the best-understood lifeform on Earth. Yet this fame is completely undeserved. *E. coli* makes up only about 0.1 percent of the bacteria in a typical human gut.

These days scientists have a much clearer picture of our inner ecosystem. We know now that there are 100 trillion microbes in a human body. The population of microbes in your own body is greater than the population of all people who have ever lived. Those microbes are growing all the time. So try to imagine for a moment producing an elephant's worth of microbes. I know it's difficult, but the fact is that in your lifetime you will actually produce five elephants' worth of microbial biomass. You are basically a microbe factory.

The microbes in your body at this moment outnumber your cells by ten to one. And they come in a huge diversity of species—somewhere in the thousands, although no one has a precise count yet. By some estimates there are 20 million microbial genes in your body. The human genome contains only 20,000 protein-coding genes. In other words, if you were to catalog all the genes in your body, 99.9 percent of them would belong to microbes. So the Human Genome Project was, at best, a nice start. If we *really* want to understand all the genes in the human body, we have a long way to go.

Now, you could say, "Who cares? They're just wee animalcules." Those wee animacules are worth caring about for many reasons. One of the most practical of those reasons is that they have a huge impact on our "own" health. Our collection of microbes—the microbiome—is like an extra organ of the human body. And while an organ like the heart has only one function, the microbiome has many.

When food comes into the gut, for example, microbes break some of it down using enzymes we cannot make ourselves. Sometimes the microbes and our own cells have an intimate volley, in which bacteria break down a molecule partway, our cells break it down some more, the bacteria break it down even more, and then finally we get something to eat.

Another thing that the microbiome does is manage the immune system. Certain species of resident bacteria, like *Bacteroides fragilis*, produce proteins that tamp down inflammation. When scientists rear mice that don't have any germs at all, they have a very difficult time developing a normal immune system. The microbiome has to tutor the immune system in how to do its job properly. It also acts like an immune system of its own, fighting off invading microbes and helping to heal wounds.

While the microbiome may be an important organ, it's a peculiar one. It's not one solid hunk of flesh. It's an ecosystem, made up of thousands of interacting species. To understand the microbiome, therefore, it helps to recall the principles that Hutchinson developed at Linsley Pond.

Even a lake as small as Linsley Pond has an impressive diversity of species. Our bodies are turning out to have huge amounts of diversity as well. Scientists can find hundreds of species of bacteria in your nose alone. And the person next to you may have some of the same species, but also have dozens of other species you lack. If you want to compare your nose to two other people's, you'll need a Venn diagram to visualize all the shared and unique species.

This sort of diversity is made possible thanks in part to the vast number of ecological niches in the human body. Microbes that live on the surface of the skin can get lots of oxygen, but they also bear the brunt of sun, wind, and cold. Microbes in the intestines have next to no oxygen, but they have a much more stable habitat. It turns out that the bugs on your fingers are different from the ones on your elbow. The two sides of a single tooth have a different diversity of microbes.

The diversity of microbes in our bodies is also generated from their intimate interdependence. In some cases, they work together to break down food. In others, one species will change the chemical conditions in our bodies to support another one. As scientists figure out these partnerships, they are now able to rear many once-unrearable species.

Such is the case for one microbe called *Synergistetes* that lives in the mouth. On its own in a petri dish, it struggles to grow. But if you add a streak of *Parvimonas micra*, it can take off. It's not clear what *P. micra* is doing for *Synergistetes*, but it's doing something really important. There are links like this between the hundreds of species in every mouth.

Hutchinson recognized that Linsley Pond was born some 11,000 years ago and matured ecologically over the following millennia. When we are born, our bodies are also empty lakes. Babies are sterile in the womb, but they are colonized even as they're being born.

There's an element of chance to how a baby's ecosystem matures. How it's delivered at birth determines the ecology of its skin. Babies delivered vaginally are coated in the bacteria that live in their mothers' birth canal. Babies born by cesarean section end up with bacteria that live on the mother's skin. But from these different starting points, our ecological succession converges on the same kind of profile. Toddlers end up with the same set of species on their skin.

The diversity of a baby's ecosystem also increases over time. And this diversity is itself an important feature of the microbiome. It makes the ecosystem more resilient, and we benefit from that resilience. Scientists demonstrated how important diversity can be when they ran an experiment on mice. They lowered the diversity of microbes in the guts of mice and then exposed them to *Salmonella*. A low diversity of healthy microbe species made it easier for *Salmonella* to take hold and grow.

Microbes ward off invaders in many ways. They can clump onto pathogens, to prevent them from burrowing into host cells. They can form an impenetrable biofilm to shut out the competition. They can make toxins. They can send signals that effectively tell invaders just to calm down. In our mouths, they even make biosurfactants. In other words, our mouths are too slippery for pathogens to take hold.

When invaders do manage to get in, our ecosystem changes. Experiments have shown that when pathogens invade a mouse's gut, the diversity of its residents drops. The effect is akin to what happened when alewives recolonized Connecticut lakes: they sent shock waves through the food webs.

Another shock to our inner ecology comes from antibiotics. Antibiotics wipe out not only the pathogens that make us sick, but a lot of the ones that make us healthy. When antibiotics work, only the beneficial bacteria grow back. But the body's ecosystem is different when it recovers, and it can remain different for months, or even years.

In the September 2010 issue of the journal *Microbiology and Molecular Biology Reviews*, a team of researchers looked over this sort of research and issued a call to doctors to rethink how they treat their patients. One of the section titles sums up their manifesto: "War No More: Human Medicine in the Age of Ecology." The authors urge doctors to think like ecologists, and to treat their patients like ecosystems.

Recall, for example, how Hutchinson saw life in a lake as a flow of energy through a food web. For microbes, the energy doesn't come from the Sun (at least not directly). It comes from the food we eat. The energy flows into the microbes, into our own cells, and, in some cases, back to the microbes again. Microbes, it turns out, are at a strategic point in that flow, where they can influence how much energy we get from our foods. As a result, scientists have found, obese mice have a different microbial ecosystem than regular mice. And if you take the stool from one of these obese mice and transplant it into a mouse that has been raised germfree, the recipient mouse will gain more weight than recipients of normal gut microbes. The microbes themselves are altering how obese mice are processing energy.

Obesity is just one medical disorder among many that the microbiome can influence. It's also been linked to inflammatory bowel disease, obesity, colon cancer, hypertension, asthma, and vascular disease. If we can manipulate our inner ecosystems, we may be able to treat some of these diseases.

Here's one crude but effective example of what this kind of ecosystem engineering might look like. In 2008 Alexander Khoruts, a gastroenterologist at the University of Minnesota, found himself in a grim dilemma. He was treating a patient who had developed a runaway infection of *Clostridium difficile* in her gut. She was having diarrhea every fifteen minutes and had lost sixty pounds, but Khoruts couldn't stop the infection with antibiotics. So he performed a stool transplant, using a small sample from the woman's husband. Just two days after the transplant, the woman had her first solid bowel movement in six months. She has been healthy ever since.

Khoruts and his colleagues later analyzed the microbes that were in the woman both before and after the transplant. Beforehand, her intestines were filled with a bizarre assortment of species, many of which are normally never seen in the gut. But soon after the stool transplant, those exotic species disappeared—along with the *C. difficile*. Her husband's microbes took over.

It's a remarkable success, but Khoruts and his colleagues can't say exactly why it worked. He doesn't know which of the hundreds of species that they delivered to the patient restored a healthy ecosystem in her gut. Eventually, scientists may come to understand the microbiome so well that they will be able to manipulate it with surgical precision, applying just a few species in a pill, or perhaps even just one keystone species.

But to know how to do that, they'll have to explore the web of interconnections in our bodies, in the same way that ecologists can see a universe in a little lake.

CARL ZIMMER is the author of twelve books about science and writes frequently for *The New York Times*. He has been writing the blog *The Loom* since 2003. Find him at http://carlzimmer.com.

ACKNOWLEDGMENTS

There were 720 submissions for this year's anthology, and we could not have winnowed the field down to the 51 final selections without the help of our volunteer reviewers. Grateful acknowledgment for their labors to Rhett Allain, Emily Anthes, Sean Carroll, Charles Q. Choi, Kate Clancy, Mo Costandi, Krystal D'Costa, Carmen Drahl, Ann Finkbeiner, Adam Frank, Pamela Gay, Greg Gbur, Jason G. Goldman, Lisa Grossman, Nicole Gugliucci, David Harris, Lee Kottner, David J. Kroll, Tom Levenson, Glendon Mellow, Josh Rosenau, Scicurious, Janet Stemwedel, Brian Switek, Andrew Thaler, Hannah Waters, Christie Wilcox, Emily Willingham, Ed Yong, Kevin Zelnio, and Stephanie Zvan.

PERMISSIONS ACKNOWLEDGMENTS

Grateful acknowledgment is made for permission to reprint the following material:

"Somata" by Aubrey J. Sanders. Copyright © 2011 by Aubrey J. Sanders. All rights reserved. Reprinted by permission of the author. First published on *Black Ink Obelisk*. http://blackinkobelisk.tumblr.com/

"Make History, Not Vitamin C" by Eva Amsen. Copyright © 2011 by Eva Amsen. All rights reserved. Reprinted by permission of the author. First published on *Expression Patterns*. http://blogs.nature.com/eva/

"Saving Ethiopia's 'Church Forests'" by T. DeLene Beeland. Copyright © 2011 by T. DeLene Beeland. All rights reserved. Reprinted by permission of the author. First published on *PLoS Guest Blog/The Wild Muse*. http://blogs.plos.org/

"What It Feels Like for a Sperm" by Aatish Bhatia. Copyright © 2011 by Aatish Bhatia. All rights reserved. Reprinted by permission of the author. First published on *Empirical Zeal*. www.empiricalzeal.com/

"Incredible Journey" by Lee Billings. Copyright © 2011 by Lee Billings. All rights reserved. Reprinted by permission of the author. First published on *Boing Boing*. http://boingboing.net/

"In the Shadows of Greatness" by Biochem Belle. Copyright © 2011 by Biochem Belle. All rights reserved. Reprinted by permission of the author. First published on *There and (Hopefully) Back Again*. http://biochembelle.wordpress.com/

"A View to a Kill in the Morning" by Deborah Blum. Copyright © 2011 by Deborah Blum. All rights reserved. Reprinted by permission of the author. First published on *Speakeasy Science*. http://blogs.plos.org/speakeasyscience/

"It's Sedimentary, My Dear Watson" by David Bressan. Copyright © 2011 by David Bressan. All rights reserved. Reprinted by permission of the author. First published on *History of Geology*. http://blogs.scientificamerican.com/history-of-geology/

"Menstruation Is Just Blood and Tissue You Ended Up Not Using" by Kate Clancy. Copyright © 2011 by Kate Clancy. All rights reserved. Reprinted by permission of the author. First published on *Context and Variation*. http://blogs.scientificamerican.com/context-and-variation/

"Joule's Jewel" by Stephen Curry. Copyright © 2011 by Stephen Curry. All rights reserved. Reprinted by permission of the author. First published on *Reciprocal Space*. http://blogs.nature.com/scurry/

"Unraveling the Fear o' the Jolly Roger" by Krystal D'Costa. Copyright © 2011 by Krystal D'Costa. All rights reserved. Reprinted by permission of the author. First published on *Anthropology in Practice.* http://blogs.scientificamerican.com/anthropology-in-practice/

"Free Science, One Paper at a Time" by David Dobbs. Copyright © 2011 by David Dobbs. All rights reserved. Reprinted by permission of the author. First published on *Neuron Culture.* www.wired.com/wiredscience/neuronculture/

"*Tinea* Speaks Up: A Fairy Tale" by Cindy M. Doran. Copyright © 2011 by Cindy M. Doran. All rights reserved. Reprinted by permission of the author. First published on *The Febrile Muse.* http://febrilemuse-infectious-disease.blogspot.com/

"Man Discovers a New Life-Form at a South African Truck Stop" by Rob Dunn. Copyright © 2011 by Rob Dunn. All rights reserved. Reprinted by permission of the author. First published on *Scientific American Guest Blog.* http://blogs.scientificamerican.com/guest-blog/

"Moon Arts: Fallen Astronaut" by Claire L. Evans. Copyright © 2011 by Claire L. Evans. All rights reserved. Reprinted by permission of the author. First published on *Universe.* http://scienceblogs.com/universe/

"Science Metaphors: Resonance" by Ann Finkbeiner. Copyright © 2011 by Ann Finkbeiner. All rights reserved. Reprinted by permission of the author. First published on *The Last Word on Nothing.* www.lastwordonnothing.com/

"Bombardier Beetles, Bee Purple, and the Sirens of the Night" by Jennifer Frazer. Copyright © 2011 by Jennifer Frazer. All rights reserved. Reprinted by permission of the author. First published on *The Artful Amoeba.* www.theartfulamoeba.com

"Mpemba's Baffling Discovery" by Greg Gbur. Copyright © 2011 by Greg Gbur. All rights reserved. Reprinted by permission of the author. First published on *Skulls in the Stars.* http://skullsinthestars.com/

"Romeo: A Lone Wolf's Tragedy in Three Acts" by Kimberly Gerson. Copyright © 2011 by Kimberly Gerson. All rights reserved. Reprinted by permission of the author. First published on *Endless Forms Most Beautiful.* http://kimberlygerson.com/

"Rats, Bees, and Brains: The Death of the 'Cognitive Map'" by Jason G. Goldman. Copyright © 2011 by Jason G. Goldman. All rights reserved. Reprinted by permission of the author. First published on *The Thoughtful Animal.* http://scienceblogs.com/thoughtfulanimal/

"Don't Panic! Sustainable Seafood and the American Outlaw" by Miriam C. Goldstein. Copyright © 2011 by Miriam C. Goldstein. All rights reserved. Reprinted by permission of the author. First published on *Deep Sea News.* http://deepseanews.com/

"On Beards, Biology, and Being a Real American" by Joe Hanson. Copyright © 2011 by Joe Hanson. All rights reserved. Reprinted by permission of the author. First published on *It's Okay to Be Smart.* www.itsokaytobesmart.com/

"I Love Gin and Tonics" by Matthew Hartings. Copyright © 2011 by Matthew Hartings. All rights reserved. Reprinted by permission of the author. First published on *Sciencegeist.* http://sciencegeist.net/

"Adorers of the Good Science of Rock-Breaking" by Dana Hunter. Copyright © 2011 by Dana Hunter. All rights reserved. Reprinted by permission of the author. First published on *En Tequila Es Verdad.* http://entequilaesverdad.blogspot.com/

"'I Feel the Percussive Roar on the Skin of My Face'" by Karen James. Copyright © 2011 by Karen James. All rights reserved. Reprinted by permission of the author. First published on *The Guardian Science Blog.* www.guardian.co.uk/science/blog

"Freedom to Riot" by Eric Michael Johnson. Copyright © 2011 by Eric Michael Johnson. All rights reserved. Reprinted by permission of the author. First published on *The Primate Diaries*. http://blogs.scientificamerican.com/primate-diaries/

"Inside the 'Black Box' of Nuclear Power Plants" by Maggie Koerth-Baker. Copyright © 2011 by Maggie Koerth-Baker. All rights reserved. Reprinted by permission of the author. First published on *Boing Boing*. http://boingboing.net/

"This Ain't Yo Momma's Muktuk" by Rebecca Kreston. Copyright © 2011 by Rebecca Kreston. All rights reserved. Reprinted by permission of the author. First published on *Body Horrors*. http://bodyhorrors.wordpress.com/

"Chessboxing Is Fighting for Good Behavior" by Andrea Kuszewski. Copyright © 2011 by Andrea Kuszewski. All rights reserved. Reprinted by permission of the author. First published on *Rogue Neuron*. www.science20.com/rogue_neuron

"Mirror Images: Twins and Identity" by David Manly. Copyright © 2011 by David Manly. All rights reserved. Reprinted by permission of the author. First published on *The Definitive Host* and *Lab Spaces*. http://davidmanlysblog.blogspot.com/; www.labspaces.net/index.php

"Did the CIA Fake a Vaccination Campaign?" by Maryn McKenna. Copyright © 2011 by Maryn McKenna. All rights reserved. Reprinted by permission of the author. First published on *Superbug*. www.superbugtheblog.com/

"Free Will and Quantum Clones" by George Musser Jr. Copyright © 2011 by George Musser Jr. All rights reserved. Reprinted by permission of the author. First published on *Scientific American Observations*. http://blogs.scientificamerican.com/observations/

"Dear Emma B" by PZ Myers. Copyright © 2011 by PZ Myers. All rights reserved. Reprinted by permission of the author. First published on *Pharyngula*. http://freethoughtblogs.com/pharyngula/

"Faster Than a Speeding Photon" by Chad Orzel. Copyright © 2011 by Chad Orzel. All rights reserved. Reprinted by permission of the author. First published on *Uncertain Principles*. http://scienceblogs.com/principles/

"Sunrise in the Garden of Dreams" by Puff the Mutant Dragon. Copyright © 2011 by Puff the Mutant Dragon. All rights reserved. Reprinted by permission of the author. First published on *Puff the Mutant Dragon*. http://puffthemutantdragon.wordpress.com/

"Volts and *Vespa*: Buzzing About Photoelectric Wasps" by John Rennie. Copyright © 2011 by John Rennie. All rights reserved. Reprinted by permission of the author. First published on *The Gleaming Retort*. http://blogs.plos.org/retort/

"How to Stop a Hurricane" by Casey Rentz. Copyright © 2011 by Casey Rentz. All rights reserved. Reprinted by permission of the author. First published on *Scientific American Guest Blog*. http://blogs.scientificamerican.com/guest-blog/

"Shakes on a Plane: Can Turbulence Kill You?" by Alex Reshanov. Copyright © 2011 by Alex Reshanov. All rights reserved. Reprinted by permission of the author. First published on *Blogus Scientificus*. http://blogusscientificus.blogspot.com/

"How Addiction Feels: The Honest Truth" by Cassie Rodenberg. Copyright © 2011 by Cassie Rodenberg. All rights reserved. Reprinted by permission of the author. First published on *The White Noise*. http://blogs.scientificamerican.com/white-noise/

"Ten Million Feet upon the Stair" by Chris Rowan. Copyright © 2011 by Chris Rowan. All rights reserved. Reprinted by permission of the author. First published on *Highly Allochthonous*. http://scienceblogs.com/highlyallochthonous/